PROBLEMS IN HISTORICAL EPISTEMOLOGY

SYNTHESE LIBRARY

STUDIES IN EPISTEMOLOGY,

LOGIC, METHODOLOGY, AND PHILOSOPHY OF SCIENCE

Managing Editor:

JAAKKO HINTIKKA, *Florida State University, Tallahassee*

Editors:

DONALD DAVIDSON, *University of California, Berkeley*
GABRIËL NUCHELMANS, *University of Leyden*
WESLEY C. SALMON, *University of Pittsburgh*

VOLUME 191

JERZY KMITA

Professor of Logic and Methodology of Science,
Adam Mickiewicz University, Poznań

PROBLEMS IN
HISTORICAL EPISTEMOLOGY

Translated from the Polish by Michael Turner

D. REIDEL PUBLISHING COMPANY

A MEMBER OF THE KLUWER ACADEMIC PUBLISHERS GROUP

DORDRECHT / BOSTON / LANCASTER / TOKYO

PWN-POLISH SCIENTIFIC PUBLISHERS

WARSZAWA

Library of Congress Cataloging-in-Publication Data

Kmita, Jerzy.

[Z problemów epistemologii historycznej. English]
Problems in historical epistemology / Jerzy Kmita ; translated from the Polish by Michael Turner.
p. cm.—(Synthese library ; v. 191)
Translation of : Z problemów epistemologii historycznej.
Iucludes index.
ISBN-13: 978-94-010-7136-9   e-ISBN-13: 978-94-009-1421-6
DOI: 10.1007/978-94-009-1421-6
1. History—Philosophy.   I. Title.
D16.8.K55513 1988   87-21918
901—dc19   CIP

This translation has been made from Z problemów epistemologii historycznej published by Państwowe Wydawnictwo Naukowe, Warszawa, 1980.

English edition published by PWN—Polish Scientific Publishers, Miodowa 10, 00-215 Warszawa, Poland, in co-edition with D. Reidel Publishing Company, P.O. Box 17, 3300 AA Dordrecht, Holland.

Distributors for U.S.A. and Canada :
Kluwer Academic Publishers, 101 Philip Drive, Norwell, MA 02061, U.S.A.

Distributors for Albania, Bulgaria, Cuba, Czechoslovakia, German Democratic Republic, Hungary, Korean People's Republic, Mongolia, People's Republic of China, Poland, Romania, the U.S.R.R., Vietnam, and Yugoslavia :

ARS POLONA, Krakowskie Przedmieście 7, 00-068 Warszawa, Poland.

Distributors for all remaining countries :
Kluwer Academic Publishers Group, P.O. Box 322, 3300 AH Dordrecht, Holland.

TABLE OF CONTENTS

PREFACE

It was only after I had finished this volume that I came across the book by Barry Barnes, *Scientific Knowledge and Sociological Theory* (Routledge and Kegan Paul). I am in full agreement with certain ideas expounded in that book, although it also contains others that I must object to. I have decided to make some remarks about them at the beginning of my book, as I believe that they may prove useful by way of introduction to the English version of this volume. I hope that anyone who has professional reasons to turn his attention to this volume will have acquainted himself with *Scientific Knowledge and Sociological Theory* before he proceeds any further.

I fully share Barnes' view that it is possible and desirable to undertake descriptive-sociological investigations of scientific research. The main subject of this research should be the natural science, and moreover, such findings in these sciences whose cognitive value has never been questioned by professionals. These investigations must avoid becoming entangled in epistemological controversies, and through epistemology in philosophical controversies. They must not defend any of the contended theses and must not literally rely on evaluative premises that have been questioned. Thus it is possible and desirable to treat science as one of the domains of culture that can be studied by sociology (or, as the author of *Scientific Knowledge and Sociological Theory* would have said, one among the complex of such domains). I especially concur with Barnes over his claim that Mannheim's sociology of knowledge does not approach science from the purely sociological point of view. (And the same is true for the phenomenological conception of sociology, I would add, though I have not enough space here to elaborate on the subject). When Mannheim speaks about the

'scientific character' of research on society, he holds that the task
of the sociologist is to explain instances of ideological distortion
of truth, as if, à la Hegel, he has already learned the truth.
Marx does the same for that matter. In Mannheim's writings
the sociologist is explicitly engaged in epistemological controversies, and through epistemology he gives an absolute status
to his own truths or the truths of the fraction to which he
belongs. Again, Marx did the same when he acted in the role
of a journalist who draws his arguments from science.

I take issue with Barnes, on the other hand, when he says
that it is unimportant for sociology of knowledge what conclusions have ben reached in philosophical analysis of science—
an analysis which is always engaged epistemologically. It is
possible that my defense of these findings is motivated by the
fact that for a long time I have been personally engaged in
such investigations, and now I am determined to defend their
honour. I do not intend, however, to use epistemological arguments in support of my position. I will emphasize two other
reasons instead.

First, I believe that one can effectively argue for the existence of what I call social methodological consciousness of
scientific practice (a counterpart, in some degree, of Durhheim's
collective consciousness). If we grant that such consciousness
exists, it is not difficult to confront various epistemological
versions of philosophy of science with corresponding historical-
-social states of that consciousness and find out that the
versions of philosophy of science represent states of the consciousness ; that they serve as 'witnesses' to the existence of
those states. But Professor Barnes makes no provisions for permitting a sociologist of science to study this role of philosophy
of science. Neither has he made any room for functional explanation of historical-social states of methodological consciousness (or states of culture represented by historically
determined scientific disciplines, in his terminology) in terms
external to consciousness. If, however, such explanation is not
forthcoming, we will be left with the idealist formula *spiritus*

flat ubi vult. Where sociological research on science is at stake, I know of no better proposal to break out of the confinement imposed by this formula than Marx' historical materialism, as long as it is properly interpreted. In the right interpretation, it permits us to pinpoint functional reasons for the production of the relevant results of scientific research and for the emergence of methodological consciousness which regulates that production and is appropriately represented by particular versions of the philosophical investigation of science.

Secondly, philosophy of science, as a type of philosophy, is entitled, I believe, to undertake epistemological or even ethical evaluation of research practice. This practice is controlled by such evaluations, and philosophy only expresses them. If such evaluations were to be discontinued, philosophy of science vould serve no purpose. This would on turn obliterate cultural motivation for engaging in scientific research, and consequently science would cease to exist as a social phenomenon.

To sum up : this book fully acknowledges that the epistemological claims of the philosophical analysis of science are groundless. It has been written, however, in an attempt to demonstrate that philosophical analysis of science is indispensable as an expression of the methodological consciousness which regulates scientific research and as an instrument of cultural-axiological (i.e. epistemological or ethical) evaluation of its results.

CHAPTER 1

EPISTEMOLOGICAL COGNITION AS HISTORICAL COGNITION*

As the present-day philosophy (or methodology, if you will) of science frees itself from the domination of the ideas advanced by logical positivists, research in the field is taking on an increasingly historical character. This trend meets with the approval of a large number of contemporary representatives of the discipline. Imre Lakatos' statement that 'methodology is wedded to history' is not an isolated opinion nowadays.

Note that for the adherent of the opinion that the methodology of science, hereafter to be called the theory of scientific knowledge (we have one of the meanings of this term in mind here—and will discuss it later on), is a historical discipline, epistemological reflection on historical research plays a special role. In this view, cuch reflection is not only a part of 'ordinary' analyses within the framework of that theory; it is naturally addressed to the theory itself. For the theory of historical cognition, from the point of view under consideration here, which is also the viewpoint accepted in this chapter, is not just a fragment of a general theory of scientific knowledge, but is also a metatheory relative to the latter. This may give rise to certain difficulties of a logical and epistemological nature, but it is probable that they can be overcome. The conception to be outlined below seems to confirm that belief.

Of course, that particular role played by the theory of historical cognition within the theory of scientific knowledge is evident in the first place to the adherents of the approach usually termed historicism. Those who oppose that approach will tend to treat it lightly. It would, however, be difficult to defend simultaneously the whole range of opinions usually

labelled 'historicism', and we shall accordingly concern ourselves with an analysis of the concept expressed by that term.

1.1. FACTOGRAPHICAL *VERSUS* THEORETICAL HISTORICISM

Whenever we speak here of historicism, we shall be using the
term in the gnosiological sense exclusively. Yet even if we
consider the term in this sense alone, it is easy to note that
the general concept expressed thereby is represented by many
specialized variations, which often oppose one another in very
essential respects. We shall try to grasp why this is so by referring to the following regulatory definition of that concept.

The name of historicism—within a given sphere of research—
will be assigned to every such approach which postulates the
use in that sphere of a definite set of methodological norms and
directives that are considered specific to historical research.

It is evident that particular varieties of the general concept
of historicism are defined (in practice usually only implicitly)
by : (i) indication of a given research discipline, (ii) designation
of a given set of methodological norms and directives which are
binding societally or are suggested individually as norms and
directives specific to historical research. If, for instance, it is
stated that a given case involves historicism in economics, linguistics, or art theory, then several different possibilities can
still come into the picture. These are linked with corresponding
sets of methodological norms and directives which do not mutually coincide. It is thus obvious that only an explicit relativization of a given use of the concept of historicism to definite
set of norms and directives makes it possible to eliminate verbal
misunderstandings.

Of course, the concept of historicism can also be used as
a general framework, rather than in a concrete way. In this
case, the relativization we are interested in naturally resembles

the sense of parametric variables. This framework concept of historicism (in a given field of research) may be rather poor in content, since it involves only one feature of the relevant gnosiological approaches, namely, that each approach requires observance (in a given field of research) of an indefinite set of methodological norms and directives treated as specific to historical research. Nevertheless, there are situations in which it is convenient to refer to that framework concept. Before we take up this issue, let us briefly outline a feature which is common to such concepts as historicism, for the use of these concepts seems particularly characteristic of the historical disciplines.

We repeat that each such concept can be applied either concretely or as a framework concept. It thus has two senses, and both these senses are expressed by means of the same term. Next to 'historicism' we can mention 'evolutionary change', 'dominant relations of production', 'truth value', 'work of art', 'linguistic expression', etc. The first meaning, here to be called detailed-historical, manifests itself when the term involved is completed by an appropriate concrete relativization, which may take on various forms, according to the nature of the concept in question. In particular, a given concept may be related to a definite set of beliefs, whether individual or observed societally (the rendering of Znaniecki's humanistic coefficient). Such a concept will then be classified as subjective[1], examples of which include historicism (relativization to a definite set of methodological norms and directives), work of art (relativization to a given set of beliefs which constitute societal or individual artistic and aesthetic consciousness), linguistic expression (relativization to an appropriate linguistic competence), etc.

The other meaning of the concepts in the group under consideration has been termed a framework one. The content of a given framework concept is not constituted by unambiguously indicated features of its designata ; that content consists of *sui generis* 'variables' (in a sense close to that used by Husserl), each of which represents any feature of a given kind, and all the configurations of features resulting from the 'filling' of those 'variables' are assumed to fulfil a common, specified condition. Thus, in particular, historicism, in its framework sense, is an approach endowed with a 'variable' feature involving reference to an indefinite set of methodological norms and directives, and this set is assumed to be specific to historical research.

This opposition : framework concept *versus* its detailed-historical analogue, is linked to a factor which in the present writer's opinion is very characteristic of the historical disciplines.

We shall return to this issue before long, but for the time being we shall be concerned with another opposition, namely that of theoretical *versus* factographical historicism. This opposition is essential for us because the thesis advanced above in a preliminary manner, i.e., the thesis that the theory of scientific knowledge is a historical discipline, has interesting consequences only when accompanied by additional assumptions which formulate with greater precision the attitude of its advocate toward historicism, and a historicism of a definite kind at that.

Thus, we can distinguish two types of historicism in the framework sense : factographical and theoretical. The factographical type is constituted by those approaches which assume, as specific to historical research, methodological norms which do not postulate the formulation of laws or which even prohibit their formulation. The first part of the above alternative expresses a general standpoint characteristic of varieties of factographical historicism : the specific nature of historical research is seen neither in the manner of substantiating laws and theories, nor in the type of such laws and theories, but in the sphere of the remaining results of research. The second part expresses an additional element : it assumes the program of idiographism. That latter variation of factographical historicism refers to methodological norms according to which, in historical research, one should in no case formulate laws. We can thus single out two versions of factographical historicism, radical and moderate.

On the other hand, theoretical historicism assumes, as a specific element of historical research, methodological norms that require the formulation of historical laws or law systems (theories), and these are to be of a specific kind and/or are to be substantiated in a specific manner.

For greater clarity it may be said briefly that

(a) moderately factographical historicism sees the specific nature of historical research outside the sphere of laws and/or theories, but does not exclude them from the set of possible results of its research ;

(b) radically factographical historicism not only sees the specific nature of historical research outside the sphere of laws and/or theories, but considers the elimination of these to be an additional feature of that research ;

(c) finally, theoretical historicism considers the manner of substantiation of laws or their systems (theories) to be a specific feature of historical research.

Of course, theoretical historicism is also represented by several variants. One of the most important differentiations might well concern the fact that laws postulated as specifically historical may have either a theoretical character or a phenomenalistic one. Yet we will not explore this matter here.

But we should not be astonished by the fact (easily to be established if concrete examples of these three varieties of historicism are selected) that one and the same variety is represented by positions often vehemently at odds with each other and which are commonly believed to be otherwise diametrically opposed to one another. One reason at least why this ought not to be surprising is that—as we already know—variants (a), (b), and (c) are all represented by diverse positions, each of which refers to a different set of methodological norms and directives, specific (from a given standpoint) to historical research. Between individual positions advocating historicism in the detailed-historical sense, there may be consi-

derable differences of outlooks, leading nevertheless to conclusions that are analogous from the perspective of interest to
us here.

Radically factographical historicism is the common deno-
minator of the conceptions which accept Dilthey's and Spran-
ger's idea of humanistic *Verstehen* and those which accept the
thesis of the Baden School, according to which the basic
concepts used by a humanistically-minded historian are
constructed by 'reference to value', and hence must be singular
concepts which are neither suitable nor intended for the construction of laws (because the facts which embody the various
values are unique). Both these groups of outlooks are in agreement on the conclusion that the result of historical research (in
the case of both viewpoints, only the humanities are of concern here) should not include laws and *a fortiori* theories, but
the grounds for this postulated state of affairs are quite different in the two cases.

Similar oppositions, perhaps even more deep-reaching ones,
divide the advocates of the various types of moderately factographical historicism. That is the meeting ground of thinkers
of varied provenance. They include W. Dray, according to whom
all arguments that historians cannot and should not formulate
laws, because the facts they are concerned with are unique,
are unconvincing.[2] He adds, however, that 'showing that there
is no metaphysical barrier to bringing historical events under
laws is not the same as showing that the laws are in fact used,
or that they are in practice available'.[3] Thus laws may occur
among the results of historical research, but this fact has no
bearing upon the specific features of that research. For the point
is that 'historical events and conditions are often unique simply in the sense of being different from others with which it
would be natural to group them under a classification term—
and different in ways which interest historians when they
come to give their explanations.'[4] When a historian is engaged
in explaining, say the course of the French Revolution, he does
not strive to explain it as just any revolution, 'as an

astronomer might be interested in explaining a certain eclipse as an instance of eclipses ; ... he might even say that his main concern will be to explain the French Revolution's taking a course unlike any other, that is to say, he will explain it as unique in the sense distinguished above'.[5]

Thus, in W. Dray's opinion the peculiarities of historical research do not arise because the historian is, for fundamental reasons, unable to formulate laws, nor because such laws would have to be idiosyncratic in some respect. These peculiarities are linked to the specificity of the manner of explanation used by the historian, explanation which only incidentally refers to laws.

W. Dray's standpoint is that of moderately factographical historicism, because, as can easily be seen,

(i) it assumes that historical research is regulated by its own specific set of methodological directives which at least concern explanation ;

(ii) these directives do not at all apply to laws, though laws are admissible among the results of research.

The author of *Laws and Explanation in History* differs from many other representatives of moderately factographical historicism by virtue of his opinion that explanation may take place without reference to laws, historical explanation being the actual realization of that possibility. Since that operation consists in the specifically historical association of facts (in several ways), usually not on a nomological level, the methodological directives concerned with historical explanation bear no relation to the procedure of formulating and substantiating of laws.

Of course, all those who exclude the possibility that explanation may assume no laws at all are inclined to argue with Dray quite vehemently on this very point. But if they notice some methodological peculiarities of historical research, though not in the type of laws arrived at nor in the manner of formulating and substantiating them, they represent, together with Dray, moderately factographical historicism. For if it is claimed (typically) that professional historians are as a rule not concerned with formulating and substantiating scientific laws, but engage in describing facts by historical statements,[6] then it is obvious that the methodological peculiarity of historical research is assumed to consist in the fact that such research concentrates on detailed factographical description. At the same time, even the very possibility of establishing specific laws within historical research is dismissed, because it has become customary for substantiated theorems of rigorous generality to be called sociological or economic laws, but not historical laws.[7] The possibility we are interested in here is certainly not excluded by the mere fact that certain terminological conventions prevail rather than others, but by the fact that these conventions are regarded as an adequate reflection of the real situation : explanation in history makes reference to laws established in psychological or sociological research. A historian may also discover laws, but not only will his formulations not differ from psychological or sociological laws, they will in fact be laws of this kind.

If we disregard the various speculative constructions, such as Toynbee's, we have to state that at the present only Marxist philosophy can aspire to the role of advocate of scientific theoretical historicism. That historicism is also marked by a fairly radical universalism : it postulates formulation and substantiation of specific historical laws not only relative to social research, but also far beyond that sphere. We shall not discuss here the intricate problem, still far from being solved, of the scope of Marxist theoretical historicism. Nor shall we try to describe in any greater detail the assumed specificity of historical laws. For the purpose of this study it is sufficient to point out—as we shall do shortly—a certain characteristic feature of these laws. Note that not only Marxist theoretical historicism may be based on the recognition of this feature. Nevertheless one of the distinguishing traits of Marxist historicism is a reliance of this feature on a methodological directive requiring that all research into social phenomena assume historical materialism as its theoretical (explanatory) foundation.

Since science, regardless of whether it is understood as a special type of social praxis or as a form of social consciousness, functionally related to the praxis, certainly belongs to the sphere of social phenomena, the theory of scientific knowledge is subject to the norm of theoretical historicism, and thus in particular ought to assume historical materialism. (It seems, moreover, at least with regard to social research, that this requirement exhausts the normative content of Marxist historicism.)

We shall hereafter be concerned with the consequences of this conclusion, which may also be expressed by reference to the title of this essay : Marxist epistemological cognition is theoretical-historicist cognition. But we shall first point to a characteristic feature of historical laws. It is interesting insofar as it helps us understand the origin of the apparent correctness of the stand adopted by the advocates of factographical historicism, radical factographical historicism (idiographism) included.

It is also to the point to note that for Karl Marx, not only is it true that all science of society should be historical in the sense of complying with the norm of theoretical historicism, but this is the only opportunity for social science to reach a theoretical level, i.e., to go beyond a registering systematization of common practical wisdom. Classical economics did make certain endeavours in that respect, but they were both incomplete and inconsistent, and vulgar economics did not even try to cross that line. This was so because bourgeois economics was marked by theoretical ahistoricism. Hence, for instance, when taking up an analysis of the monetary form of exchange value we face a task 'the performance of which has never yet even been attempted by bourgeois economy, the task of tracing the genesis of this money-form, of developing the expression of value implied in the value-relation of commodities, from its simplest almost imperceptible outline, to the dazzling money-form'.[8]

1.2. FRAMEWORK REGULARITIES

The thesis in this second part of the essay is that historical laws in a large sense (including laws of the theory of biological evolution) invariably strive to give an account of framework regularities. In other words, they have the form of nomological formulas. Let us begin with examples.

The following phenomenon has been explained. In a population of the butterfly species *Basilaria archippus*, a certain set of wing coloration traits $t_1, \ldots, t_n$ achieves dominance. The explanans involved consists of two statements:

(i) The set of traits $t_1, \ldots, t_n$ emerged earlier by mutation. It also marked the population of *Danais plexippus*, more numerous and threatened by the same enemies (birds). Yet the *Danais plexipus* population frightens birds away thanks to a poisonous substance in the organisms of its members. This set of traits and set of circumstances are such that owing to them, the population of *Basilaria archippus* is better adapted.

(ii) Whatever the set of traits that emerge in any population by mutation and whatever the circumstances in which that population lives, if the type of those traits and those circumstances is such that the population in question is better adapted, then that set of traits under those circumstances becomes common in the population.

Now part (ii) of this explanans, which by the way records one of the more important laws of the theory of natural selection, is an example of a nomological formula and the regularity it describes (on the assumption that it does hold) is an example of a framework regularity. We would like to bring out three characteristic features of nomological formulas.

First, what a nomological formula predicates about variables of the lowest order (in our case the variable of the lowest order is that which ranges over the set of all populations of species) are not constants, but variables too. In our example no definite traits that are to emerge in a population by mutation are mentioned, nor are the circumstances in which that population lives. Both elements are expressed by variables which range over two corresponding sets of traits of any population. Certain constants are predicated only about those variables of a higher order (e.g., the three-place relation of better adaptation that holds among the set of unspecified mutational traits, the set of unspecified circumstances, and a population).

Second, those variables of a higher order cannot, for practical and theoretical reasons, be replaced by molecular predicates in the form of finite disjunctions. For practical reasons, because there are too many various combinations of a concrete set of mutational characteristics with a concrete set of natural circumstances of a given place and time. For theoretical reasons, because we do not know the laws on the basis of which it would be possible to predict all the possible mutations within a given population, and all the possible natural circumstances in which a given population might live. Thus concrete substitutions for higher-order variables in nomological formulas must be made by an analysis of given situations in which the fact to be explained occurs.

Third, we have to note that when transforming a nomological formula into a 'law' that includes no higher-order variables and when doing so by taking into account the appropriate concrete data which characterize the fact to be explained, we arrive at a general statement, but one which is as a rule satisfied non-vacuously by one case only, namely the fact being explained. For instance, the nomological formula discussed above would become a statement which says that every species in which the concrete set of traits $t_1, \ldots, t_n$ would emerge by mutation and which would co-exist with another, more numerous, population, etc., would after some time reveal the domination of the said set of traits. There seems to be no doubt that this statement could be satisfied non-vacuously only by the population of *Basilaria archippus*.

Let us consider one more example. In my opinion, the particular laws of historical materialism are nomological formulas or even sets of such formulas. For instance, the statement that the social existence of human beings determines their consciousness would have to be expressed by at least two nomological formulas. One of them, which corresponds in a purely formal way to the above-discussed principle of natural selection would be roughly as follows.

Whenever in a social class K, determined by its place in a given socio-economic structure, there emerges a new set of beliefs B, which in a subjectively rational manner motivates a praxis which is more effective, relative to the actual objective needs, than the praxis motivated in a subjectively rational manner by the set of beliefs B' so far socially accepted in K, the set of beliefs B will in the course of time drive out the set of beliefs B' in K.

Moreover I think that the fundamental general statements of the social sciences cultivated on the basis of historical materialism are nomological formulas, too. In particular, therefore, not only do interstructural regularities (which characterize a number of socio-economic structures) have the character of a framework, but so do regularities characterizing singular structures.

In summary of this part of the discussion, we may state that the assumption under which historical laws are characterized, among other things, by the fact that they report framework regularities makes it possible to identify one of the rational sources of the beliefs of the advocates of factographical historicism, and radical factographical historicism (idiographism) in particular. For the point is that when trying to explain a given fact by a nomological formula, we must analyse in great detail the situational context of the fact, because the appropriate repertory of predicative constants is not given to us in advance. If we tried to reconstruct, on the basis of a given act of historical explanation, a law constructed by analogy to a typical law of physics, we would obtain a statement that would be satisfied nonvacuously only by the situation being explained. Hence there is something pertinent in the saying that the physicist studies recurrent phenomena, whereas the historian stu-

dies unique phenomena. Of course, the formulation is misleading : every individual phenomenon is unique, and in that respect there is no difference between a single event explained by
a physicist and one explained by a historian. The difference is
that a representative of a mathematized natural discipline is
interested in a given single fact as a typical specimen of a definite homogeneous class of facts that satisfy a given physical
or chemical law. He explains it in the same way in which he
would explain any other single fact in that class. On the contrary, a historian who establishes, by his analysis, that a given
fact occurs in the context of a certain, essentially unique, combination of its (otherwise general) properties or relations, can
explain that fact not by that combination alone, but only by
referring to its definite type. In other words, in the former
case an individual fact is explained by pointing out that it
belongs to the class of facts determined by the combination of
the circumstances $c_1, \ldots, c_n$; in the latter, by pointing out that
it belongs to the class of facts determined by a definite type of
combination of circumstances that cannot be predicted in advance : each combination corresponds in principle to one fact
only, and some such combination occurred in the situation under consideration. The set-theoretical aspect of this opposition
could be shown in a symbolic form as follows (where a stands
for a given individual event ; ω is a variable that ranges over
the set of such events ; C_i stands for a given circumstance
treated as a property of an event ; X_i is a variable ranging
over the set of such circumstances ; T stands for the type of
a given combination of circumstances, i.e., for the condition
satisfied by those circumstances) :

(1) $$a \in \{\omega: C_1(\omega) \wedge \ldots \wedge C_n(\omega)\},$$

(2) $$a \in \{\omega: \bigvee_{X_1, \ldots, X_n} |X_1(\omega) \wedge \ldots \wedge X_n(\omega) \wedge T(X_1, \ldots, X_n)|\},$$

where the product of every combination $X_1, \ldots, X_n$ of circumstances of type T is as a rule a unit set.

As can be seen, explaining a given fact by means of a nomological formula means subsuming it under a certain framework concept, in the earlier described sense of that term.

Note also that the adoption of the view that a characteristic feature of historical laws is that they establish framework regularities would make it possible for the advocates of theoretical historicism, including Marxists, to formulate their opinions with greater precision. It is sometimes the case that the difference between their views and those of the advocates of factographical historicism is difficult if not impossible to grasp. For instance, S. Kozyr-Kowalski and J. Ładosz write that:

'Materialistic historicism does not question ... the possibility of constructing research apparatus by singling out in the social process that which is always present, recurrent, and universal, but it stresses that a theory that attempted to explain social reality with the help of such an apparatus would not be in a position properly to assess the role of universal phenomena in the social process, because such phenomena always occur in the setting of empirical reality and condition that reality in its historical forms. In the case of ahistoricism, it is forgotten that there are no universal phenomena outside history, it separates theoretically that which is universal from that which is historical, failing to comprehend that that which is universal can exist only as that which is historical; it endows the product of its theoretical abstraction with independent existence, makes it an independent element of the historical process to which it gives historically defined forms. The materialistic nature of Marxist historicism is manifested, among other things, in the fact that it does not accept the posibility of the existence of universal phenomena outside a definite historical form. For it, all auniversal phenomena are simultaneously historical phenomena.'[9]

It does not appear that the above statement could be understood as a (somewhat obscure) manifestation of the standpoint of Marxist theoretical historicism otherwise than on the assumption that it protests against explaining historical facts by including them in classes of events determined by universally recurrent circumstances. For should it protest also against establishing framework regularities, it would be difficult to say in what the opinions formulated above differ from the standpoint of W. Dray, according to whom in historical research laws can in principle be established, but are not useful for that research.

1.3. ASSUMPTIONS OF HISTORICAL EPISTEMOLOGY

The thesis that the theory of scientific knowledge is a historical discipline, when interpreted in the spirit of Marxist theoretical historicism, determined that historical materialism should form the explanatory basis of that theory. But even this leaves many detailed issues to be settled concerning the way to pursue Marxist epistemology. To present concisely certain fundamental (in the present writer's opinion) conditions for research into that epistemology, we shall list some statements which are : abbreviated definitions (which require further expansion and more precise formulation), or special applications of assumptions of historical materialism, or specified solutions of issues mentioned above, or consequences of statements of these three kinds.

0.1. A dynamic functional structure relative to a property W is a relational system which continuously preserves W and whose elements or sets of elements take on such states from their corresponding repertories that without their taking on such states the system would lose W ; that property W is dynamic in nature ; its concrete form manifests itself in various ways in various periods, but it always satisfies a definite framework condition.

0.2. If in a functional structure relative to the property W, we can single out at least one substructure which is also functional, relative to a property W', and if the preserving of W' by that substructure is necessary for the preserving of W by the whole structure, then such a structure is termed a hierarchical functional structure.[10]

0.3. The states of a kind R of an element (set of elements) of a functional structure relative to W, such that without states of the kind R the structure would lose W, are said to be a (more or less adequate) answer to the structure's demands ; the statement that a given state of an element (set of elements) is of the said kind R is equivalent to the statement that the function of a given element (set of elements) consists in taking on states of the kind R, and to the statement that the structure 'demands' a state of the kind R to be taken on by a given element or set of elements of the structure.

0.4. If in a hierarchical functional structure we can single out two substructures S' and S'' which are functional structures relative to properties W' and W'', respectively, and such that the preserving by S' of W' is an answer to the demand of S'', and if the preserving by S'' of W'' is in turn an answer to the demand of the whole structure, then S' is said to be functionally subordinated to S''.

1.1. Social praxis consists of the totality of subjectively-rational human actions which together with the accompanying system of objective socio-economic conditions (determined by the nature of relations of production and the level of development of the productive forces) forms a dynamic hierarchical functional structure relative to the property of reproducing existing objective conditions and/or producing conditions of a new type, with the provision that :

(i) the various types of social praxis, or independent sections of the social division of labour, form functional substructures which are directly or indirectly subordinated to that substructure consisting in the fundamental 'material' praxis, namely production and exchange ;

(ii) every type of social praxis includes a successive substructure : its socio-subjective context, i.e., the set of beliefs whose observance by individual participants in social praxis of a given type is a necessary condition for its providing adequate answers to the demands addressed to it by social praxis as a whole.

1.2. The socio-subjective context of social praxis of a given type will be called a form of social consciousness. Observance of beliefs associated with a corresponding form of social consciousness by an individual participant in social praxis is a necessary condition for the attainment of individual goals connected with the various actions undertaken on the subjectively-rational basis.

1.3. The development of social consciousness is determined functionally and genetically; social acceptance of a new set of beliefs, i.e., its emergence in social consciousness, is explained by two factors: the functional one: the new set of beliefs provides a more adequate answer to existing or arising demands in the context of actual objective conditions than did the previous set of beliefs; the genetic one: the new set of beliefs is a definite transformation of received 'mental material' (in the sense used by F. Engels).

2.1. In the sphere of social consciousness we can single out two kinds of beliefs: (a) those which combine to form the content of social practical knowledge, (b) those which combine to form the theoretical or *Weltanschauung's* complementation of social practical knowledge. Beliefs of kind (a) are generalizations which define the way to realize immediate practical goals and express current knowledge (common practical wisdom) or scientific practical knowledge (as a rule characterized by laboratory experimentation). Beliefs of kind (b) are the sum of two not disjoint sets of beliefs, one of which includes elements of socially accepted scientific theories, and the other, beliefs which determine values superior to immediate practical goals and establish relationships between those values and immediate practical goals.

2.2. The function of scientific praxis is to provide the appropriate practical sectors with predictive premises (permitting the effects of activity to be predicted). This is realized through the codification and deductive systematization of definite elements of social practical knowledge. This is a framework characterization of that function. It is realized in various ways, in

various periods and in various spheres of social scientific praxis. We can, in particular, single out the period when social
scientific praxis responds to the social demand that 'calls it to
life'. This response involves recording, codifying and systematizing elements of the received common practical wisdom.
This is the pre-theoretical period. In the next stage, the theoretical stage, science formulates theories which are (logically)
incomparable with the common practical wisdom.[11] It does so
by making an increasing use of scientific (experimental) practical knowledge where the ordinary recording of common practical wisdom ceases to be an adequate response to the demands
of social praxis addressed to research.

2.3. The subjective social context of scientific praxis takes
on the form of social methodological consciousness, consisting
of methodological norms and directives. Verbalization of social
methodological consciousness is the concern of the theory of
scientific knowledge, which always represents a certain philosophical orientation.

2.4. Those theories of knowledge which have arisen in history
as a rule contain only the normative part, which verbalizes
social methodological consciousness (usually in a fragmentary
way); their social acceptance depends on the effectiveness with
which they perform at least one of their two functions: that
of a subjective social regulator of scientific praxis (at a given
stage of development), and the world-outlook function.

2.5. Social research praxis, and hence also social methodological consciousness, change as social praxis as a whole develops,
since they respond to demands submitted by the latter. This
is why the relation between two successive systems of scientific knowledge may be of two kinds. The systems may function in a subjective social context of two different methodological-theoretic viewpoints, being thus incomparable with each
other: they are semantically interpreted in two different domains (i.e., they have different intended or standard semantic
models), in other words, they have different literal (semantic)
references. But they may also function with the same methodo-

logical-theoretic humanistic coefficient, and then their confrontation on the ('internal') logical level is possible.

Described with maximum brevity, these are some major assumptions of historical epistemology conceived in the spirit of Marxist theoretical historicism. Historical epistemology, from this point of view, is the theoretical history of the development of scientific praxis, in particular its subjective social context, verbalized more or less adequately by the various theories of knowledge. We could also use here the term methodology of the sciences in a specified sense, but this term has many other (often unconsciously assumed) meanings: it is sometimes used to denote social methodological consciousness and is also sometimes applied to its philosophical verbalization (whether spontaneously or deliberately). If this verbalization is accompanied by its axiological acceptance, then that term has the same extensions as the phrase 'normative part of the theory of scientific knowledge' suggested above. In view of this ambiguity of the term methodology of the sciences, preference is given here to the term theory of scientific knowledge.

We should like now to adduce two fairly important arguments in favour of the superiority of Marxist historical epistemology to other theories of scientific knowledge. These arguments show the possibilities, closed to the other theories, for Marxist historical epistemology to solve the problems of (i) determinants of the development of science, (ii) the relationship between historical epistemological statements and established gnosiological norms.

The first problem, that of the determinants of scientific development, today attracts the attention of researchers who cultivate the theory of scientific knowledge. The present situation of research conducted by representatives of non-Marxist trends can be characterized as follows. On the one hand, it is assumed that the development of science is a consequence of researchers respecting certain historically constant methodological norms and directives. This point of view now has its principal advocate in 'orthodox' hypotheticism. In accordance with the standpoint still taken by K. R. Popper in his *Logik der Forschung*, that trend of thought is interested—unlike logical empiricism (logical positivism)—not so much in the conditions that must be satisfied by a given system of knowledge to be scientifically valid, as in the logical relation (in the broad sense of the term) that must hold between such a system and its successor if the latter is legitimately to drive out its predecessor. The solution is offered by Popper's principle of rational criticism, accepted axiologically and also assumed to be an adequate reconstruction of the suprahistorical invariant of social methodological consciousness: the successor system should be corroborated by those data which falsify its predecessor and (the addition due to I. Lakatos) should have a corroborative surplus relative to its predecessor.

On the other hand, this hypothetistic conception, which, as we have seen, puts to the fore the issue of the principles that rule the development of scientific knowledge, and thus opposes logical positivism, which interprets the theory of scientific knowledge in a totally ahistorical manner, has been opposed by numerous epistemologists who (sometimes proceeding, moreover, from other assumptions of hypotheticism) adduce numerous arguments, based on a detailed analysis of the history of science, which seem to indicate that the principle of rational criticism does not, and even cannot, play any history-making role. The opponents, primarily P. K. Feyerabend, T. S. Kuhn, and N. R. Hanson, point to the fact that the occurrence of pieces of evidence which refute a scientific theory has not, in the history of science, always resulted in the rejection of such a theory. Moreover, according to P. K. Feyerabend, the same empirical 'raw data' can be articulated by two different theories in two different ways, by two different sets of basic statements, so that each set confirms 'its own theory'. To make matters worse, neither the principle of rational criticism nor any other principle that refers to the possibility of a logical comparison of various systems of knowledge can be operative in the case of 'scientific revolutions', because the systems involved are not comparable logically. This is why the opponents of hypotheticism have come to the conclusion that we are not in a position to indicate any general principle of the development of science ; this applies not only to the methodological principle (whose observance would explain that development), but to any principle whatsoever. They do formulate partial explanations of various kinds, but on the whole the formula stating that *spiritus flat ubi vult* prevails. Thus, for instance, the domination of Kuhn's paradigm explains a long perseverance of a certain humanistic coefficient in the sphere of methodology and theory, characteristic of the results of research (that coefficient can, by the way, be treated as an analogue of Kuhn's paradigm), but both the reason why a certain paradigm comes to dominate and the reason why it collapses remain in fact a mystery. Neither the fact that at a certain moment sufficiently many researchers come to approve (for various individual reasons) the paradigm which emerges in this way nor the fact that at another moment sufficiently many researchers come to pay attention to the anomalies that had been disregarded so far, have any explanatory force in the eyes of an advocate of theoretical historicism.

Thus, the first of the two standpoints outlined above does not notice or does not want to notice the actual logical incomparability of systems of knowledge with different methodological-theoretic humanistic coefficients, but raises to the role of a suprahistorical principle the methodological norms it accepts ; its actual observance in research is claimed to be able to explain the development of scientific knowledge.[12] On the other hand, the second standpoint notices this incomparability, but for that very reason its representatives are inclined to pursue epistemology only in the spirit of (radically) factographical historicism.

Note that both these positions are linked by a certain common element, which is very important from the Marxist point of view : both remain within the sphere of 'historical ideology' in the Engelsian sense of the term. This is to say, both interpret the development of science in a manner which does not go beyond the methodological-theoretic humanistic coefficient of the individual systems of knowledge, and hence in a socially subjective manner. Further, that coefficient is interpreted ahistorically and is therefore non-temporally applied to all systems of knowledge or treated as a series of transient paradigms. But to be able to compare systems of knowledge which are logically incomparable because of their different coefficients one must go beyond that coefficient and discover its objective determinants.

We shall refer here to the assumptions of historical epistemology presented above. It has been assumed that every system of scientific knowledge (in particular, every empirical theory) has its literal reference, i.e., its semantic model relativized to the methodological-theoretic humanistic coefficient which actually holds with respect to that system. But every such system, if it functions as an element of social consciousness, has a reference of another kind, which might be called the praxis-objective reference (of a given system of knowledge). The latter is identical with the domain of objective conditions of that sphere of social praxis the socio-subjective context of which includes social practical knowledge, codified and systematized by a given system of scientific knowledge. From the point of of view of the assumptions made here, the following relation holds between a given system of scientific knowledge and the praxis-objective reference of that knowledge : the fact that a given system of scientific knowledge corresponds more adequately than any other system to demands of social praxis in the sphere of the system's praxis-objective reference, i.e., the fact that that system relatively most effectively rationalizes a given sphere of a social praxis (which makes that praxis itself respond more adequately to the demands 'addressed' to it) explains the fact of social acceptance of that system of knowledge.

By expanding this statement we will obtain a formulation which might well characterize a certain framework regularity

in the development of social scientific praxis ; this will herein
be called the principle of the development of science.

Note above all that when a system S of scientific knowledge
is replaced by a system S', it never happens that S' represents
an absolutely perfect knowledge, i.e., that its literal reference
is identical with its praxis-objective reference. But for S to be
replaced by S' it suffices that the latter, by its literal reference, represents a better (or at least as accurate an) approximation to the praxis-objective reference of S than does (as does)
the literal reference of S ; and that S', by contrast to system S,
takes into account some new elements of social practical knowledge. For in such a situation, S' responds more adequately than
S to the demands of social praxis taking place within the
domain of its praxis-objective reference.

Hence the principle of the development of science suggested
here is as follows.

The following state of affairs is a sufficient condition for
the social acceptance of the system S' of research results in
lieu of the previously socially accepted system S of research
results : (i) the literal reference of S' is at least as adequate an
approximation of the praxis-objective reference of S as the
literal reference of S is ; (ii) S' deductively codifies and systematizes new elements of social practical knowledge in contrast to those codified and systematized by S.

The claim that this framework regularity[13] really takes place
in the history of science will prove much more convincing if
we consider how that regularity manifests itself in research
praxis.

To do so we single out two different basic situations. In the
first case, the methodological-theoretic humanistic coefficient
of S and S' is the same. Then the literal reference of S is a subdomain of the literal reference of S' or the two domains are
identical. Then also S is explained by S', which is at most
a certain generalization of the former. At the same time it is
obvious that the corresponding subdomain of the literal reference of S' is as adequate an approximation of the praxis-ob-

jective reference of S as is the literal reference of S itself. But since new elements of social experience are taken into account by S' (see (ii) above), S is replaced by S'. As is usually said in methodological studies, S' explains everything that S does and explains some new facts as well.

In the second case, S and S' have different methodological--theoretic humanistic coefficients and are thus logically incomparable. In such a situation it may be possible to demonstrate that S', by virtue of its literal reference, represents a more adequate approximation to the praxis-objective reference of S than does the literal reference of the latter. Full demonstration of the fact that that is possible would require—in the light of the assumption of historical empistemology adopted here—a functional and genetic explanation of the social acceptance of S ; we would have to show that the literal reference of S' is identical with the praxis-objective reference of S, and that the converse explanation is not possible. The most essential point of that explanation would consist in demonstrating that the domain of social praxis rationalized by S (through a deductive codification and systematization of the appropriate sector of social practical knowledge) is, in the context of the objective conditions interpreted by S', sufficiently effective. That fact would primarily explain the social acceptance of S.

Of course, full application of that procedure would be possible only with the deliberate acceptance of the assumption of Marxist historical epistemology, suggested here, but in the history of science, in situations of the kind analysed above, scientists undertake research operations which represent such essential elements of the explanation procedure described here that they can be treated as a specific, enthymematic demonstration of the fact that S', which succeeds S, by its literal reference represents a more adequate approximation to the praxis-objective reference of S than the latter does. For instance, an answer is provided to the question of how far one can effectively use the Ptolemaic theory in practice under the assumption that the corresponding domain ob objective conditions of that practice (e.g., irrigation of fields, navigation, etc.) is characterized by the heliocentric theory; likewise, an answer is provided to the question of how far one can effectively use Aristotelian physics in practice under the assumption that the objective conditions of that practice are characterized by classical physics. Also Marx answers the question of how far a capitalist producer can effectively use the assumptions drawn from vulgar economics, assuming his own theory of the capitalist model of production.

The second condition for social acceptance of S' in lieu of S consists in taking into account new elements of social experience; the situation analysed here (wherein the coefficients of S and S' differ) primarily involves scientific practical (laboratory-experimental) knowledge. But it is characteristic that in this situation, the second condition is much less important than in the first situation, wherein S and S' have the same coefficient. The general theory of relativity, for instance, was preliminarily accepted even before it could claim new experimental data.

Research operations undertaken in the context of the second situation usually are said, in physics and in the methodology of physics, to satisfy the principle of correspondence. That principle has various formulations. It will be interpreted here as part of the principle of the development of science described under (1). It could accordingly be worded thus:

It is a necessary condition for the social acceptance of a system S' of research results in lieu of the previously socially accepted system S of research results that the literal reference of S' be at least as adequate an approximation of the praxis--objective reference of S as that which is offered by the literal reference of S.

With respect to the two situations analysed above, in which the relevant system S and S' may find themselves, one might speak of explanatory (cumulative) correspondence (case one) and strict correspondence (case two).[14] Hereafter in these essays, however, another pair of terms will be used to designate these two cases of correspondence, namely : 'generalizing correspondence', and 'essentially corrective correspondence'. As we shall see shortly (in the next chapter), colloquially functional semantic intuitions that arise in connection with these terms more effectively suggest the sense herein intended.

By making use of the principle of correspondence thus interpreted, we can formulate another principle, which considers the principle of the development of science in another way. It is as follows :

It is a necessary condition for the social acceptance of a system S of research results that it correspond directly or indirectly (through the correspondence-based intermediary of other systems) with social practical knowledge.

This will be termed the principle of empiricism[15].

1.4. THE RELATION BORN BY GENERAL STATEMENTS OF HISTORICAL EPISTEMOLOGY TO METHODOLOGICAL NORMS AND DIRECTIVES

It has been demonstrated in the preceding section that neither the occurrence of 'scientific revolutions' nor the logical incomparability of successive stages in the development of scientific knowledge push historical epistemology to a position of radically factographical historicism. But to avoid such a situation, it is vital that we rely on assumptions of historical materialism in order to leave the sphere of 'historical ideology'. We shall now consider another argument in favour of Marxist historical epistemology, which shows that a specific positioning of methodological norms and directives within that epistemology removes a problem which has long troubled theorists of knowledge.

It can easily be seen that the principles formulated above have counterparts in certain methodological directives. Thus we may speak, in particular, about the directive of correspondence or that of empiricism. The former would state that in order to arrive at scientifically valid empirical knowledge we have to construct a given system of knowledge in a manner that ensures its correspondence with the preceding system, while the latter would say that in order to arrive at scientifically valid empirical knowledge we have to construct a given system of knowledge in a manner that ensures its direct or indirect correspondence link with social practical knowledge.

What is the relationship between those directives and the respective principles ?

Note first that both principles have been formulated with reference to the concept of socially accepted (empirical) knowledge, and the directives outlined above refer to the concept of scientifically valid (empirical) knowledge. Now the former concept is beyond all doubt relativized in a historical and social manner, which need not be the case when it comes to the latter concept. Moreover, at first glance it might seem that the social and historical relativization of the concept of scientifically valid knowledge deprives it of the status of an evaluative concept ; hence, if the two methodological directives we are interested in here were to assume that relativization, they would bear no relation to the epistemological norm that requires accumulation of scientifically valid knowledge. We shall now try to counter this notion by defending the opinion that :

(a) the social and historical relativization of the concept of scientifically valid knowledge need not deprive it of its defined cognitive-axiological sense ;

(b) that concept, redefined in a certain respect and relativized in the above manner, is—in view of its descriptive sense—co--extensional with the (rather) non-evaluative concept of socially accepted knowledge.

Let us begin with (b). We shall first treat the concept of scientific validity as a socially subjective one, i.e., relativized to the social methodological consciousness of science (more precisely : of a given field of social scientific praxis) in a given period. The concept thus denotes (for a given relativization) the set of those research results which meet the criteria corresponding to historically determined social methodological consciousness. But since social methodological consciousness is the socially subjective context of scientific praxis (in a given field), i.e., is functionally subordinated to the latter, it selects those and only those research results which effectively rationalize the corresponding domain of social extrascientific praxis. And it is just those and only those research results which are socially accepted. Thus scientifically valid knowledge (under a given relativization) covers all those research results which are socially accepted (under the same relativization). Furthermore, they are those research results which, in accordance with the established tradition of Marxist philosophy, are called relative truth or are acknowledged as historically correct (under the same relativization).

Yet, returning to point (a), why are we to think that the said co-extensionality of the concepts of socially accepted knowledge and scientifically valid knowledge does not deprive the latter of its cognitive-axiological sense ? It could be claimed that this sense of the concept in question is guaranteed only from the point of view of a historically determined social methodological consciousness, but not from the point of view of the epistemologist who studies that consciousness. Let us note that the cognitive validity of a given system of knowledge indicates, in accordance with what has been said earlier, the attainment by that system of a better (or at least equally adequate) approximation to the objective conditions of that domain of social praxis which it rationalizes, by contrast to the historically preceding system. This is guaranteed by the correspondence link between the two systems. It can be claimed accordingly that the concept of correspondence link denotes the same relation as does the concept of advance in knowledge. And the latter has an unquestionable cognitive-axiological meaning. We can thus adduce cognitive-axiological arguments in favour of observance of such methodological directives as the directive of correspondence or that of empiricim. Observance of these directives is necessary for a relative advance in knowledge.

But are such arguments not superfluous ? If the principles which correspond to those directives correctly report the regularities which in fact govern the development of science, then the demands of that domain of social praxis, and in the last analysis, the demands of social praxis as a whole (beginning with production) sooner or later enforce observance of those directives in particular. This is a special case of a problem which is well known in the Marxist literature. There are no reasons to discuss it here. By referring, on this particular issue, to the standpoint usually adopted by Marxists we can just state that the methodological norms and directives, which form the normative part of Marxist historical epistemology, are related to its corresponding statements in such a way that they define those research operations which : (a) accelerate advance in knowledge, (b) eliminate disturbances of various kinds which would slow those advances, (c) are effective in this respect in a given stage of development of a given discipline.

Of course, the instructiveness and usefulness of methodological norms and directives in Marxist historical epistemology increase as these norms and directives are formulated on the basis of less general principles, which—unlike the principles of empiricism and correspondence—are satisfied not in every period of the evolution of research practice, but only at a given stage of development of science. Thus, for instance, the directive of empiricism used by Marx in *Capital* clearly pertains to the theoretical stage of development of scientific knowledge : it requires strict, and not just any, correspondence relative to common practical wisdom. Likewise, his theory of the capitalistic mode of production does not include the generalizations assumed by the class of capitalist producers[16], (and codified and systematized by vulgar economics), but explains their social acceptance in that class by pointing to the causes of relative effectiveness of the capitalist practice based on those assumptions. This is also the case with the directive of abstraction and concretization ; for reasons which are too intricate to be discussed here briefly, that directive can be effectively applied only at the theoretical stage of the development of science. Requiring the observance of the Marxian directive of empiricism or that of abstraction and concretization at the pre-theoretical stage of the development of social scientific practice would be as reasonable as requiring the observance of norms of socialist humanism in the period of primitive accumulation or as expecting formation of capitalist enterprises under the slave system.

The assumption that the scientific validity of statements reduces to their representing relative advances in knowledge, and that this latter concept, in light of its descriptive meaning, can be applied, on the basis of the assumptions of historical materialism, to a specific domain of social development, that is, to the development of social scientific praxis delivers epistemology from what might be called a theoretical stalemate. Epistemology is driven into that stalemate by the following reasoning. The epistemologist is supposed to answer (at least among other things) the question of what kind of knowledge is valid. He accordingly proposes a criterion for such knowledge, seeing to it that he does not make the mistake of *petitio principii* (as phenomenologists, for instance, so energetically insist), that is, that the criterion not be based on existing scientific findings, because before such a criterion is established, he cannot know whether those findings are valid. But such a criterion, to be denoted here by K, must itself be substantiated and this can be done only by means of another criterion (a criterion of validity for the criterion of validity of scientific knowledge). Thus we find ourselves in a situation in which there are three options): (i) *regressus ad infinitum*, which practically amounts to total scepticism ; (ii) dogmatic adoption of the criterion K ; (iii) dismissal of the problem, which however makes it impossible to define the axiological sense of the activity of epistemologists. Note that the large majority of epistemologists choose the third solution, thus acquiescing, consciously or not, to the fact that their formulation of epistemological norms and directives does not have a defined sense. It is also

to be pointed out that many of these same epistemologists reduce the entirety of epistemologic work to this very action
(finding a criterion for validity). The second solution is also
chosen quite often : this has been done by logical positivists, but
not by them alone. Incidentally, many philosophers try to weak-
en the impression of dogmatism by basing that criterion—contrary to the positivist tradition—on ontological assumptions.
The schema is as follows : since reality is such and such in its
essence, its valid knowledge should have such and such characteristics. Of course, that pre-scientific ontology is as dogmatic as
pre-scientific epistemology.

The fact that Marxist historical epistemology does not face
the necessity of making the choice as described above is primarily due to

(1) its giving a historical interpretation to the concept of
scientific validity,

(2) its imparting to that concept a definitive descriptive meaning within the general theory of social development which
historical materialism constitutes[17].

As we have striven to show, our concept maintains its cognitive-axiological sense in this context.

NOTES

* It is to be born in mind that 'cognition' refers here not to psychological processes proper to individuals, but to a social process : the growth
and development of society's systems of knowledge (trans. note).

[1] For the description of subjective concepts see K. Zamiara, *Metodolo-
giczne znaczenie sporu o status poznawczy teorii* (*The Methodological
Significance of the Controversy over the Epistemological Status of
Theory*), Warszawa, 1974, pp. 141–143.

[2] Cf. W. Dray, *Laws and Explanation in History*, Oxford, 1957.

[3] *Ibid.*, p. 46.

[4] *Ibid.*, p. 47.

[5] *Ibid.*, p. 47.

[6] Cf. A. Malewski, 'Zagadnienie idiograficzności historii', (The Problem
of the Idiographic Nature of History), in his *O nowy kształt nauk spo-*

lecznych. Pisma zebrane (Toward a New Shape of the Social Sciences, Collected Works), Warszawa, 1975, p. 162.
[7] *Ibid*, p. 163.
[8] K. Marx, *Capital*, Vol. 1, Moscow (n.d.), pp. 44–48.
[9] S. Kozyr-Kowalski, J. Ładosz, *Dialektyka i społeczeństwo. Wstęp do materializmu historycznego (Dialectics and Society. An Introduction to Historical Materialism)*, Warszawa, 1972, p. 146.
[10] For a detailed analysis of this concept, cf. K. Zamiara, *op. cit.*
[11] The use of the concepts of social practical knowledge and of the pre-theoretical (positivist) and theoretical (post-positivist) stages in the development of science refers to the results obtained by A. Pałubicka in her book *Orientacje epistemologiczne a rozwój nauki (Epistemological Orientations and the Development of Science)*, Poznań, 1977.

It might be added here that in a few of my previous statements on the subject of the relation of theoretic scientific knowledge to common practical wisdom (or of the inter-relation of successive, qualitatively different stages of theoretical knowledge development), I have spoken either of incomparability or of incompatibility, and not simply of incomparability. This second alternative will now be eliminated, since on the plane of statements alone (and not of their references), only incomparability comes into play.

[12] It is characteristic that this standpoint is adopted by some representatives of Marxist philosophy, who are not inclined to pursue the theory of scientific knowledge in accordance with the assumptions of Marxist theoretical historicism. They stress, e.g., 'the striving for the truth', 'the principle of negation of negation', etc., as 'laws' in the development of science. The first of these examples shows clearly that in their opinion a certain (by the way, very vague) norm determines the development of that sphere of social practice which is science. This opinion is obviously at variance with historical materialism, and with respect to its generality, it is inferior to the similar approach of Popper. The second example relates to one of the laws of dialectics. But the relation of the negation of one system of knowledge to another system of knowledge is a subjective, 'mental' one, and hence using it to describe the mechanism of the development of science is a manifestation of historical ideology (in the sense used by Engels) as long as it is not demonstrated that the negation of one development stage of scientific knowledge by the succeeding stage is a secondary consequence of the functioning of the regularities that govern the entire objective process of social development.

[13] This regularity is not fundamental in character. It is a consequence of the objective function of scientific praxis previously alluded to. For a broader discussion of this question, see J. Kmita's *Szkice z teorii*

poznania naukowego (Essays on the Theory of Scientific Knowledge), Warszawa, 1976.

[14] These terms have been used by the present author up to now in, among others works, *Szkice z teorii poznania naukowego (Essays on the Theory of Scientific Knowledge)*.

[15] Of course, this principle differs essentially from the traditional, non-Marxist interpretations of empiricism.

[16] The picture which they give of socio-economic facts under capitalism was, as is known, called 'appearance' by Marx.

[17] Obviously, like any other scientific theory, historical materialism does not require any additional philosophical substantiation. No theory in physics, for instance, requires for its substantiation prior proof of an ontological thesis of the existence of physical objects.

CHAPTER 2

THE RELATION OF CORRESPONDENCE

The notion of correspondence, presented in the preceding chapter, will now be examined somewhat more closely. We will in particular be concerned with the notion of essentially corrective (strict) correspondence. For this plays an especially vital role, as we shall see, in historical epistemology. Yet a more thorough understanding of this notion requires that it be confronted with the meanings of the term 'correspondence' which are usually operative in methodological thought. Furthermore, taking these meanings into account permits an understanding of why many philosophers, with P. K. Feyerabend in the forefront, so forcefully deny that any principle (directive) of correspondence (in the traditional understanding of the word) can claim universal validity in the history of science. Mathematized natural sciences are of primary relevance here, for the notion of correspondence is mainly used with reference to the links that are supposed to connect particular developmental stages within these fields of knowledge.

2.1. LITERAL REFERENCE

The proposed analysis of the notion of essentially corrective (strict) correspondence will begin with a wider presentation of an ancillary category closely connected with correspondence— the literal reference of a given statement or set (system) of statements. This category was introduced in the preceding chapter, though without a deep analysis.

The category mentioned denotes—when it is used in reference to a definite system of statements—a particular case of what, in logic, is called the semantic model of the system. Literal reference, thus considered, is therefore a certain relational system—in other words, a certain domain (or relational structure in R. Wójcicki's sense)—with a defined universe, equipped with a series of relations defined on that universe. This set may include unary relations, from an extensional point of view, properties constituting subsets of the universe. This set, as we know, is called the characteristic of the given relational system (or domain). We say of this system or domain that it is a semantic model of a given set of statements if and only if all these statements are semantically interpreted within it[1].

The notion of semantic (and in particular, literal) reference also applies to a 'single' statement. In that case, it does not consist of an appropriate relational system (or appropriate domain), but of an ordered n-tuplet, which is to be thought of as an element of a defined relation. We will call this a state of affairs. Relations of various kind are involved here, depending on the syntactic structure of a given statement. Thus, for instance, in the case of an atomic sentence with a predicate of m arguments ('poznań is a city', 'Poznań is a smaller city than Warsaw', etc.) we are dealing with a relation of the belonging of a given m-tuplet of entities (it is possible that $m = 1$) to a given relation (possibly unary). The ordered n-tuplet, which serves here as the state of affairs, in this case contains $m+1$ positions. The reference of the sentence 'Poznań is a city' will therefore be a state of affairs of the form : ⟨denotation of the predicate : '... is a city', denotation of the singular term : 'Poznań'⟩, whereas in the case of the sentence 'Poznań is smaller than Warsaw', we will be dealing with a state of affairs of the form : ⟨denotation of the predicate : '... is smaller than...', denotation of the singular term : 'Poznań' ; denotation of the singular term : 'Warsaw'⟩. The first position of the ordered n-tuplet is associated with the denotation determined for a given predicate. A set-theoretical relation of belonging to holds between the predicate's denotation and the following m-tuplet of denotations of singular terms. In the case of sentences of different syntactic structure, the corresponding relations which are to contain given states of affairs appear with appropriate variation. This is a technical matter, however, which will not be examined here thoroughly[2].

The literal reference of a given statement or system of statements is a name—as we have seen—for a specific instance of semantic reference. It is called 'literal' in order to emphasize one of its special properties, on the basis of which it is distinguished from other possible semantic references of the given statement or statement system.

Clearly, in dealing with a given language not interpreted semantically, we can associate its primitive terms with their denotations in a multitude of ways. Denotations can be selected from various relational system (various domains[3]) so that each of these relational system (each domain) becomes a semantic model of the corresponding set of sentences (system of statements) of the given language, assuming, of course, that this set is not self-contradictory.

Nevertheless, in formulating the sentences, in constructing a given statement system, one intends at the same time to speak of a definite domain and not just of any domain which might happen to be a semantic model of a given set of sentences (statement system). This very 'target' model is what in logic is called the intended or standard model (intended or standard domain). The criterion for affirming that a given domain or semantic model is intended, i.e., standard, and hence that a standard semantic interpretation has been achieved, has no psychological character, but rather is purely semantic. Namely, the mark that a given domain or model is standard is the fact that certain singled out sentences of the corresponding language are true in it. Thus, for instance, if a formalized theory containing a predicate which we have intended to interpret as a predicate of identity is submitted to semantic interpretation, then we say that the predicate has received a standard interpretation if there has been associated with it a denotation fulfilling the axioms of identity theory, that is, a denotation with regard to which these axioms are true sentences[4].

The notion of 'standardness' of semantic interpretation is of great service to epistemological research. Consider, after all, the fact that in formulating certain statements—whether in the framework of informal discourse or of scholarly research—we always relate them to a domain defined identically for everyone ; we confer upon the constituent expressions universally established denotations. At particular instances, we need not realize this at all. It is nevertheless the case that these references are standard—in this very sense that, in formulating certain statements, we are also ready to recognize the truth of some other statements with which the notions appearing therein are implicity intertwined.

Various 'constraints' determine that particular individuals are more or less scrupulous in considering the truth value of respective elements of knowledge of the world. Among the more important factors in this regard is the fact that the social praxis of communication *via* language would be nonfunctional, if a certain 'portion' of knowledge of the world were not presumed to be true, that is, were it not socially accepted (under the pressure of objective functions of the social praxis of language communication). Thanks to this, intersubjective, standard references of relevant linguistic expressions are upheld. If, for instance, we can communicate with each other in English, using the word 'city', it is because, in using this word, we confer a standard semantic interpretation upon it, a standard denotation. Whatever the set of objects with which it is to be identified, this denotation is such that the truth of certain statements must be recognized in its light. Such statements include : 'Every city is located in a definite, geographical place', 'Every city is inhabited by a significant number of people', 'Every community that constitutes the population of a city is under a separate administration', etc. Of course, individuals can (and do) alter the standard denotations of particular linguistic expressions. They can fail to take the truth of appropriate sentences into account. However, if this operation is not in some way condoned by the society (and thus appropriately justified), sanctions threaten these individuals. The most poignant of these sanctions would surely involve the impossibility of linguistic communication in situations especially relevant to their elementary 'survival' interests (the purchase of food and clothing, the performance of occupational duties, etc.).

Parenthetically speaking, social consciousness can be thought of as the set of all those beliefs which (as a necessary condition) have to be recognized by the general public in order for social praxis to be able to function, that is, to answer objective needs. This notion of social consciousness authorizes the following assertion : if the recognition of a given belief by individuals of a particular society is semantically indispensable for linguistic communication within that society, then this belief belongs to the sphere of social consciousness.

It is these standard references of individual statements and statement systems as well as of individual notions (our concern here is simply with their denotations), defined by the pre-supposition of the truth of their corresponding sentence networks that are here called literal references. The term lends emphasis to the intuition that an epistemologist considering the relationship of these statements or of the beliefs expressed by them to reality understood in this or that light must always reckon with the possibility that that which these statements assert, their literal 'contents', need not be at all identical with the reality which may be 'hiding' behind them, whatever that reality might be. The literal reference of particular statements or of their system is a socio-subjective conception of reality ; it is an expression of reality from the viewpoint of the corresponding underlying network of statements, which are presumed to be true, i.e., are socially accepted.

The phrases 'literal references of statements' and 'literal reference of beliefs' are used interchangeably in this book. Nevertheless, it is to be stressed that the beliefs with which we are here concerned are not to be understood in the traditional psychological sense. They are rather a variety of propositions in the logical sense. For both those beliefs, the social acceptance of which qualifies the literal references of certain statements, and the beliefs expressed by the statements, as well as the references themselves, as a rule, unavailable to the conscious minds of individuals, although that which is present to an individual's consciousness does constitute a kind of idiosyncratically distorted 'trace' of these beliefs. The social consciousness comprising the beliefs determines, in varying degrees, individual consciousness. Yet this determination almost never manifests itself as a mere repetition.

From the standpoint of historical epistemology it is clear, first of all, that the literal reference of systems of scientific statements, which are of primary interest to us, cannot be identified with the network of objecive conditions which result in the social acceptance of these systems. Secondly, the historical variability of these conditions explains the variability of literal references of successive statement systems and demands that this variability be foreseen. A stand incompatible with either the first or the second outlook would also be incompatible with fundamental theses of historical materialism. To assume identity of objective reality and the socio-subjective (in particular, scientific) image of that reality is to take a radically idealist stance[5]. On the other hand, to assume that society's scientific consciousness is independent of objective conditions is to negate the over-all value of the thesis of the determination of social consciousness through the obejctive conditions of social existence. It would also be incompatible with numerous data from detailed historical investigations into the development of science.

2.2. THE CHARACTERISTICS OF ESSENTIALLY CORRECTIVE (STRICT) CORRESPONDENCE

The nework of statements (beliefs) which qualify the literal reference of a given set of research results (and in particular, of a scientific theory) will herein be called the methodological-

theoretic humanistic coefficient of that set. This designation
has been selected in order to emphasize two key points which
it incorporates. First of all, 'humanistic coefficient' is a term
introduced by F. Znaniecki. He used that term mainly to express the intuition that objects of research in the humanities
are not 'objects in themselves'. Rather, they are always somewhat 'seen' by definite subjects and in this 'perceived' form
undergo research. For obvious reasons, the present author
agrees neither that the concern of the humanities is to be restricted to objects equipped with explicitly marked 'humanistic coefficients' nor that this 'coefficient', in principle, is to
be a rendering of the consciousness of individuals. By identifying this term with a defined set of beliefs, we nevertheless take
advantage of its association with the intuition mentioned above
to stress that literal references are, in their own way, 'socially
perceived' fragments of reality. Secondly, the qualification
'methodological-theoretic' is meant to show that the given statement network, qualifying the literal reference of a given set
of research results, is composed of elements validated from the
viewpoint of social methodological consciousness—of the appropriate field of scientific research and of the appropriate historical period.

Notice that it is only under the assumption that the methodological-theoretic humanistic coefficient of a given scientific
discipline undergoes no historical changes (that, as a result,
in science—throughout its history—we 'talk' continually of the
same things), that each developmental change, leading from
one system of scientific statements S to a succeeding S', can
be treated as the result of an appropriate logical (in a wide
sense of the word) operation performed on S and yielding S'.
This may be an operation of generalization, an operation of this
or that kind of logical modification, or even an operation of
negation. Yet it is only when two given systems, S and S', literally 'speak' of the same fundamental things that it makes
any sense, or is indeed at all possible, for logical operations to
lead from one system to the other. Undoubtedly, all conceptions

treating correspondence between consecutive sets of research results as—exclusively— a kind of logical and 'internal' relation, i.e., a relation holding 'within' a mental frame, 'within' a particular methodological-theoretic humanistic coefficient, more or less (usually, less) consciously assume this historical invariability of the coefficient and of science's literal reference.

It has already been pointed out that the assumption under consideration here excludes science—whether as a type of social praxis or as a form of social consciousness—beyond the operative range of the key premises of historical materialism. Let it be added that the mere thought that the idea of a logical operation (or group of logical operations) could constitute a principle determinant of the development of science is also incompatible with historical materialism, and therefore historical epistemology, which is founded theron. Yet such a thought is assumed by the identification of all correspondence with some version of 'internal' correspondence.[6] Of course, historical epistemology concedes a relative autonomy to social scientific praxis and, in particular, to the methodological-theoretic humanistic coefficient, which constitutes an integral part of science's socio-subjective context. This coefficient may remain intact for fairly long periods, and successive changes within character. In this case, correspondence also has an 'internal' character. In this case, correspondence also has an 'internal' character and is, as a rule, a generalizing (explanatory) correspondence, i.e., the literal reference of an earlier set of research results forms a subdomain of the literal reference of a succeeding set.

The fallacy in every outlook on correspondence that portrays it as always an 'internal' relation—in other words, the fallacy in the assumption of the historical invariability of the methodological-theoretic humanistic coefficient of all successive sets of research results (of scientific theories)—can nevertheless be quite easily shown. In particular, the results of a series of the in-depth historical analyses of P. K. Feyerabend or T. S. Kuhn can be used to this end. Yet a very simple example (and, thus,

an effective one from a didactic perspective) will be adduced
here.

Both Galileo and Newton make use of the notion of gravity.
Yet is this really but one notion—with one and the same literal
reference, delimited by one and the same methodological-theoretic humanistic coefficient? If the answer to this question
were to be negative, then it would be proper to speak here
rather of two different notions, related to each other in a specific way. (We will return to this matter shortly.) In such an
event, it would also be necessary to abandon the opinion that
correspondence of an 'internal' type holds between Galileo's
theorems formulated with the help of the notion of gravity and
the corresponding theorems of Newton.

Yet, as a matter of fact, the difference here is a deep one.
First of all, gravity according to Galileo (the literal denotation
of the notion) is a relation comprising a set of ordered pairs of
the type ⟨Earth, any physical object⟩, whereas gravity according to Newton (the literal denotation of this notion) is a relation
comprising a set of ordered pairs of any physical objects. Secondly, Galileo's gravity is an anti-symmetric relation (only
the Earth attracts), whereas Newton's gravity is a symmetric
relation. Finally, gravity according to Galileo is a constant
quantity (every instance of gravity is associated with the same
numeric measure: 981 cm/sec^2), whereas gravity according to
Newton is a variable quantity (its numeric measure is functionally dependent on the product of the numerical value of the
mass of the two given objects and the square of the numerical
value of the distance between them): If the difference boiled
down to just this first point, there would be no impediment to
speaking of an 'internal', generalizing correspondence. The
remaining differences rule out such an interpretation however. Thus, only essentially corrective (strict) correspondence
can come into reckoning here.

Before undertaking a direct analysis of the notion of essentially corrective correspondence, there is a matter to which our
attention must be turned. A conclusion has just been set forth

regarding the logical incomparability of theorems built upon the notion of gravity in Galileo's sense and theorems built upon Newton's notion of gravity. Yet that conclusion is based on a comparison of the literal denotations of these two notions. Might one not object that there is an inconsistency here ? Such an objection would not be well-founded. For it is not the respective theorems which have been compared here, but rather the denotations of notions. Of these notions it has been assumed that they represent definite set-theoretical, ontological categories and that this constitutes their only common feature (comprising their plane (dimension) of comparability). Also in set-theoretic terminology, note has been taken of the differences between them. Of course, it follows from this line of reasoning that the literal references of individual statements or of their systems are in any event set-theoretical entities. This assumption can be justified and is indispensable as well. Without accepting it, the very notion of (literal) reference could not even be introduced. Nor, for that matter, could any other semantic notions.

In the preceding essay, essentially corrective correspondence was defined as a special case of correspondence in general. Correspondence is a relation between two consecutive systems of research results (scientific theories) S and S' such that the literal reference of S' comprises at least as adequate an approximation of the praxis-objective reference of S as that which is offered by the literal reference of S. The special case of essentially corrective correspondence is distinctive in that, in its context, S and S' are equipped with differing methodological--theoretic humanistic coefficients. We will now endeavour to transform this definition in such a way as to imbue it with greater (than in the preceding chapter) practical efficiency.

Let us begin by asserting that instead of the notion of some absolute praxis-objective reference of system (theory) S appearing in the definition of correspondence quoted above, the notion of a relative praxis-objective reference of system (theory)

S can alternatively be applied. We will justify this assertion, relying on the following terminological convention.

The relative praxis-objective reference of system (theory) S, relativized with respect to the system (theory) S', is the name we will give to that subdomain of the literal reference of system S' which comprises the range of practically effective applications of system (theory) S, defined, naturally, in terms of system S'.

The terminological convention just formulated defines as a relative (with respect to system S') praxis-oriented objective reference of system (theory) S such a subdomain of the literal reference of system (theory) S' that, when practical activity which employs the theorems of S in the role of predictive premises is considered from the viewpoint of S' and is located in that subdomain of S', it can be affirmed that this activity produces results sufficiently close to predictions obtainable on the basis of system S'. Sometimes the results may even be identical to these predictions. The idea of correspondence can now be characterized as follows : correspondence holds between any two systems (theories) S and S' if and only if a relative, praxis-oriented objective reference of S can be defined in terms of S'. It is relativized to S' in the sense that is constitutes a subdomain of the literal reference of S'.

If system S', corresponding in the above sense with S, is associated with the same methodological-theoretic coefficient as is system S', then it consequently differs from system S at most in that the literal reference of system S comprises a proper subdomain of the literal reference of system S'. In this case, operating within the confines of that subdomain and using the theorems of S as premises for prediction, we predict, on the basis of S, effects precisely duplicating the predictions provided by system S'. We are then dealing with correspondence of the 'internal' type : generalizing or explanatory (system S can be explained by means of system S')[7].

On the other hand, in the case where system (theory) S' corresponds with S and the two systems (theories) possess distinct methodological-theoretic humanistic coefficients, the subdomain of the literal reference of S' which comprises the range of the effective applicability in practice of the theorems of S (expressed in term of S', of course) is different from the literal reference of S. Then we are dealing with essentially corrective (strict) correspondence. In this situation, predictions which are based on the predictive premises of S and which subjectively direct corresponding practical activity differ, as a rule, (though often to an insignificant degree—when it comes to activity located by S' in the relevant subdomain of its literal reference), from predictions attainable on the basis of S'.

It can now be shown that correspondence in the sense developed in the previous essay holds if and only if correspondence holds in the sense we have just ascribed to it. They are thus ideas with the same denotation. This fact is obvious in the case of correspondence in its generalizing (explanatory) variant. For the literal reference of the correspondent[8] system (theory) is—by both definitions of this variant—at most a proper subdomain of the literal reference of the corresponding system (theory). We are then dealing—by both definitions—with a situation in which the literal reference of S' comprises at least as adequate an approximation to the praxis-objective reference of system S as does the literal reference of system S. However, as far as the essentially corrective variant of correspondence is concerned, we note that the corresponding system (theory) S', both when it is defined in terms of an absolute praxis-objective reference and when it is defined in terms of a relative praxis--objective reference, by indicating a subdomain of its own literal reference as a range of practically effective applications of S differing from the literal reference of S, makes it possible for social acceptance of S to be explained. This explanation stems from a thesis derived from historical materialism that predictive knowledge serving effective practice in a given field is subject to social acceptance. It is this very explanatory potential which constitutes the guarantee that correspondence defined—whether in the sense of the previous chapter or in this new sense—as essentially corrective leads to a system (theory) S' which provides a better approximation to the (absolute) praxis-objective reference than does the correspondent system (theory) S.

In summary, essentially corrective correspondence holds between system of research results (theory) S' and system (theory) S when S' assigns as a relative praxis-objective reference of system S a subdomain of its own literal reference which, in view of the dissimilarity of the methodological-theoretic humanistic coefficient associated with the two systems (theories), differs from the literal reference of S. This subdomain—the range of practically effective applications of system (theory) S, characterized in terms of S'—comprises, as has been pointed out, an essential correction of the literal image of praxis-objective reality provided by S.

In the particular case of our previous example, gravity in the sense of Newton is in a relation of essentially corrective correspondence to gravity in Galileo's sense. To be precise, this correspondence holds between Newton's theorems concerning gravity and the corresponding theorems of Galileo. That is, the praxis-objective reference relativized to the Newtonian system of Galileo's theorems about gravity consists of those cases of Newton's gravity which concern the earth and some physical object of relatively small mass located fairly close to the surface of the earth. By treating the relevant theorems of Galileo as though they referred in effect to this particular case of Newtonian gravity, we achieve on their basis predictions very similar to predictions attainable on the basis of Newton's theorems[9].

Notice also that the above example can fairly easily be presented in such a way as to create the impression that the gravity of Galileo is a special case of Newtonian gravity. Parenthetically speaking, that is the very way in which this example is regularly presented in physics. As a matter of fact, all cases of essentially corrective correspondence, as a rule, are presented in this way in physics. The range of effective applicability in practice of theory T, expressed in terms of theory T', is presented as though it were the literal reference of T. This is a symptom—as well as frequently a cause—of essentially corrective correspondence being ahistorically conceived of as 'internal' generalizing correspondence. We will now take up this very problem.

2.3. REMARKS ON TRADITIONAL UNDERSTANDINGS OF CORRESPONDENCE

This presentation of traditional understandings of correspondence will begin with a short reflection upon the relationship between two statements characterized by the following traits.

(1) They belong to different systems of knowledge. The literal reference of the first system is designated by methodological-theoretic humanistic coefficient C, whereas the literal reference of the second is designated by coefficient C'.

(2) The second system of knowledge is in a relation of essentially corrective correspondence to the first.

(3) Each of the statements asserts there to be a link joining an action of specific type along with its accompanying circumstances to a predicted and intended effect (the goal of the action). Hereafter, to simplify our formulations, we will speak merely of an action, assuming that this notion also entails the accompanying circumstances.

Consequently, we are dealing, on the one hand, with a statement which claims that very action characterized—in light of C—as an action of kind A leads (without fail) to an effect characterized—in light of C—as an effect of kind E and on the other hand, with a statement which claims that every action characterized—in light of C'—as an action of kind A' leads (without fail) to an effect characterized—in light of C'—as an effect of kind E'. It may be, in a particular case, that action of type A and action of type A' involve the bringing about of states of affairs within the context of which certain parameters are assigned numerical values p and p' (e.g., to a given object, there is assigned a specific velocity). Functionally linked (in the mathematical sense) with these are states of affairs of types E and E', within the context of which certain other parameters are assigned numerical values q and q' (e.g., transit of a certain distance by an object of defined velocity in defined time span). In this case, the two statements assert their respective functional dependences (e.g., the functional dependence of the numerical measure of time of transit of given distance on numerical measure of velocity). It may also be, moreover, that only the connection between A' and E' has the character of functional dependence.

(4) A' is the correspondence rendering of A, and E' is the correspondence rendering of E. We will also say, in such a situation, that A and A' as well as E and E' are correspondence renderings of each other. What does this mean ? A deeper analysis of this idea will be presented in the next chapter. Here, however, we merely point out that, in the situation under discussion, it can be affirmed from the viewpoint of C' that the idea A consists of a conceptualization, framed by C, of the class of states of affairs of A' (the class denoted in the context of C' by the idea A'). Each state of affairs indicated by A' is conceived of in the context of C as a state of affairs of kind A. Thus, for instance, if within the context of the methodological--theoretic humanistic coefficient proper to Newton's notion of gravity, it is ascertained that the action of the 'letting go' of some physical object has been carried out, an action of kind A', at the same time, anyone who 'looks' at the action from the viewpoint of the coefficient characteristic of Galileo's notion of gravity speaks of it as an action of kind A, producing the initial conditions for the operation of the Galilean law of free fall, not for the Newtonian law of general gravitation.

Analogical observations are relevant to effects of type E and type E' as well.

It is worth noting here that the difference between correspondence renderings A and A' as well as between E and E' seems enormously difficult to grasp due to the fact that usually the first and the second action (the first and the second effect) are both denoted by the same linguistic expression. (For instance, Galileo's gravity and Newton's gravity are both denoted with the help of the same word : 'gravity'.)

We shall say of a statement which belongs to set of research results S' equipped with methodological-theoretic humanistic coefficient C', and which asserts a dependence (in particular, a functional dependence) of E' on A' that it is a simple correspondence translation in terms of S' (or onto S') of a statement asserting in the context of methodological-theoretic humanistic coefficient C, a dependence (possibly functional) of E on A if and only if A' is a correspondence rendering of A and E' is a correspondence rendering of E. In the event that, in the context of S', an appropriate dependence can only be claimed for E' on some subrange of A'—and after all, this is the situation with which we are normally involved—we will say of that assertion (of dependence) which can be accepted within the framework of S' that it is the corrected correspondence translation of the respective assertion of dependence of E on A. The notion of correspondence translation (in general) will denote for us the sum of the denotations of the ideas we have just introduced.

As can be easily observed, the literal reference of a simple or corrected correspondence translation comprises the relative (with respect to S') praxis-objective reference of the translated statement.

We should also note that the situation we've been considering, that is, the situation in which the dependence of E on A and the dependence of E' on (an appropriate sub-range of) A' are asserted within the framework of isolated statements constitutes only one of the variants of a more general picture. The remaining variants are represented by cases in which we are involved with extensive, internally ordered conjunctions. The first component of every such conjunction concerns action A or A', the final concerns effect E or E'. To take all the remaining variants into account, however, would complicate the course of our exposition immeasurably, but without yielding anything essentially new from the point of view of interest to us. For this reason, we will confine ourselves to considering the simplest case.

Let us turn now to traditional understandings of correspondence. Because they have an 'internal' character, which can now also be expressed by the assertion that they presume the synonymy, without exception, of notions A and E with their correspondence renderings A' and E', the idea of correspondence translation introduced above can therefore be conveniently generalized so that it will also refer to individual cases of 'internal' correspondence. Thus, I hereby propose the generalization : from now on we shall also speak of correspondence renderings of A and E in the form A' and E' when the humanistic coefficient C' differs in no way from coefficient C, hence the correspondence renderings are really synonyms of each other. With this terminological convention, a simple correspondence translation may be synonymous with the translated statement, while a corrected correspondence translation may be equivalent to a restricted sense of the translated statement. It is to be stressed that not all advocates of the 'internal' concept of correspondence take that possibility into account. This question will be discussed shortly.

A special term will now come in handy for designating a correspondence translation obtained in the modality of essentially corrective correspondence. In such a case, we will speak of an essentially corrective correspondence translation.

Let it once again be emphasized that whether 'internal' correspondence may arise in certain cases is in no way being questioned here. It arises, as we have seen, when the methodological-theoretic humanistic coefficient of two appropriate, successive systems (theories) remains unchanged. We assert only that in many cases of key importance for the development of science, the coefficient does in fact undergo change and that 'internal' correspondence clearly does not then come into play. Nevertheless, certain instances of this kind of correspondence,

singled out by various researchers, can be more or less adequately characterized as variants of the 'internal' correspondence relation. It is thus permissible to take them into account,
but only upon examination of those situations in which the relation of 'internal' correspondence truly comes into play.

It has been proposed in this essay that correspondence of
the 'internal' type be termed generalizing (or explanatory).
Now, in referring to traditional understandings of this relation,
I would like to develop this proposition somewhat and, at the
same time, to provide it with a wider justification.

In can be asserted that the proposed, traditional characteristics of correspondence, pretenders to universality, although in
fact they are applicable only to appropriate instances of 'internal' correspondence, are divisible into two fundamental
groups. Within the first group, the relation of generalization
sensu stricto is involved, within the second group we find
approximate generalization. This means that either (in the case
of the first group) a correspondence translation is treated as
a synonym of the correspondent theorem, and generalized in
the framework of the corresponding system (in the framework
of the corresponding theory) or (in the case of the second
group) the correspondence translation, which is—analogously—
a particular instance of a more general theorem of the corresponding system (theory), formulates a dependence only approximately equivalent to the dependence formulated by the correspondent theorem. Of course, the synonymy of correspondence renderings A and A' as well as E and E' is still assumed.

Let us begin our review with the first group. In this context,
correspondence is sometimes spoken of globally. That is, the
nature of the theorems of system S, with which the theorems
of system S' are to correspond, is not qualified with respect to
the dichotomy : 'observational' knowledge $vs.$ 'theoretic' knowledge. Sometimes correspondence is also discussed with this
qualification.

An example of the simplest global conception is the statement asserting that correspondence is a relation in the context

of which laws governing a certain domain of physical phenomena, thanks to the establishment of new, more general laws,
become special instances of these new laws[10]. It is clear that
the correspondence translations of theorems concerning the
dependence of E on A, identified semantically—as is done in
all concepts of correspondence from the first group—with those
theorems, are treated here as simple correspondence translations in the above sense of the term : as though they were
ready-made theorems of the corresponding system S' without
restricting the range of A' as the correspondence rendering
of A. Certain other theorems of S' are to be the generalizations
of these correspondence translations-theorems. However, more
typical variants of this global conception take the fact of the
respective restriction of the range of A' into account and therefore assume that only corrected correspondence translation
comes into play.

Thus, for instance, Z. Augustynek asserts, 'Let $\langle T', T \rangle$ designate a pair, where T' is the new theory, and T is the old.
Then the basis for a relation of correspondence between them
is that for certain physical conditions V, the logical implication
$T \wedge T' \rightarrow T$ holds. The conditions V are such that when they arise,
a universal physical constant (or its inverse), appearing in T'
and characteristic of T', can be zeroed. Of course, conditions V
vary for different physical constants. Hence, it is said that
for V, theory T is the limiting case of theory T', or that theory
T' is more general (has a wider range of applicability) than
theory T. In the opinion of physicists, the relation of correspondence holds between the elements of the following pairs of
theories : $\langle STR, CM \rangle$ (zeroing of $1/c$, where c is the speed of
light), $\langle GTR, STR \rangle$ (zeroing of gravitational constant γ),
$\langle QM, CM \rangle$ (zeroing of Planck's constant h), $\langle RQM, QM \rangle$ (zeroing of $1/c$), $\langle RQM, STR \rangle$ (zeroing of h). Here CM, STR, GTR,
QM, and RQM stand respectively for : classical mechanics, special theory of relativity, general theory of relativity, quantum
mechanics and relativistic quantum mechanics'.[11]

It should be stated that the above passage very adequately

expresses the opinion of physicists therein recalled—from at least three standpoints. First of all, it enumerates those cases of correspondence which are deemed by physicists to be truly the most momentous for modern physics. Secondly, it also conceives of them—in accordance with the opinion of physicists— as cases of 'internal' correspondence, even though all of them are examples of essentially corrective correspondence. Finally, also in accordance with the opinion of physicists, the theorems of theory T (and in particular, their correspondence translations) are special instances of theorems of theory T' in the sense that they are 'limiting cases', that is, they proceed from the theorems of T' given the additional assumption of conditions V, which arise with the 'zeroing' of relevant physical constants; the conditional sentence : $V \wedge T' \rightarrow T$ is thus supposed to be a theorem of T'.

Of course, the conditional sentence above can really be counted as theorem of T' only in the event of 'internal' corresponence. When T and T' appear in the context of different methodological-theoretic humanistic coefficients, they literally 'speak' of wholly different domains and no logical connection holds between them. Moreover, in this situation, correspondence renderings from T and T' often possess different grammatical categories—e.g., one-argument predicate, shall we say, 'mass of object $x = p$' vs. two-argument predicate 'mass of object $x = p$ in view of reference system u'. In that case, our conditional sentence is even grammatically senseless. When we assume, however, that 'internal' correspondence holds here, and that T is only a correspondence translation (in terms of T') of the given correspondent theory, then our conditional sentence may be treated as a theorem of T' under one other condition. What is this additional condition ? It is that T can not be a simple correspondence translation of the given correspondent theory, but must be a corrected translation. The quoted author implicitly assumes this, since he says that the given 'logical implication' holds under 'certain physical conditions V'. This can only mean that the range of generality of the theorems of the translation is limited precisely to conditions V. When these observations are related to individual assertions of dependence of E on A and of E' on A' (the quoted passage speaks of whole theories), it can be said that we are dealing with a corrected correspondence translation in the context of which an eventuality of type E' is rendered dependent upon those eventualities of type A' which arise under conditions V. Furthermore, in accordance with the general assumption of traditional understandings of correspondence, E' is supposed to be synonymous with E, as is A' with A. Also, in accordance with a general assumption of those traditional understandings which have been singled out as the first group, the corrected correspondence translation is synonymous with an appropriately restricted statement of dependence of E upon A (restricted to conditions V). In this way, the theory symbolized by T' turns out to be a generalization of theory T (or—in our understanding—its appropriate corrected correspondence translation), because it formulates more general dependences which are to hold under all conditions not only under 'limiting' conditions V. (We might add that, upon taking the above reservations into consideration, our conditional sentence can be more strongly restated as an equivalence : what theory T' states holds and conditions V are satisfied if and only if what the corrected correspondence translation of T states also holds.)

Incidentally, the example of an 'internal', strictly generalizing conception of correspondence that has just been presented may evoke—even in its modified (in accordance with the remarks made above) form—yet another doubt. (This doubt has been raised by, among others, P. K. Feyerabend.) This concerns the purely logical correctness of the conjunction of the given theory T' and the description of 'limiting' conditions V. For this description is supposed to assert that the relevant quantity 'zeroes'. However, if theory T' asserts that the quantity is a constant, or that its absolute numerical value is never less than some positive number (as is the case with, among others, the inverse of the speed of light : $1/c$), then our conjunction is contradictory, and any set of sentences can therefore result from it (and not just theory T). Thus, for instance, the conjunction of the special theory of relativity with the assertion that the inverse of the speed of light 'zeroes' (i.e., the speed of light verges on infinity), which is supposed to yield classical mechanics (upon restriction of the generality of its theorems to situations which involve just this 'zeroing'), has a contradictory character (since STR asserts in one of its most important theorems that the speed of light is constant). This question has been brought up by way of digression, however, and we will not consider it in greater depth.

In a sketchier and yet very clear way, W. Mejbaum presents the strictly generalizing conception of correspondence in his essay 'Prawa i sformułowania' ('Laws and Formulations')[12]. He does not treat the restriction of the 'correspondent theorem' (the corrected correspondence translation—in our terminology) as the result of taking only 'limiting' conditions into account, but rather as an appropriate restriction of whatever sort.

'Physical dependence $Z'(x)$', we read in the essay cited above, 'stands in a relation of correspondence with physical dependence $Z(x)$ if :

(a) there exists a set of objects S such that, under defined physical conditions V

(α) $$\bigwedge_x \left(x \in S \rightarrow x \in \text{False } (V, Z) \right),$$

(β) $$\bigwedge_x \left(x \in S \rightarrow \sim x \in \text{False } (V, Z') \right),$$

(b) there exists a physical parameter A and a value a of that parameter such that the sentence $A(x, a) \rightarrow (Z'(x) \equiv Z(x))$ is logically true'.[13]

'False (V, Z)' ('False (V, Z')') symbolizes here the set of those objects which, in satisfying conditions V, falsify the statement of dependence Z (and Z', respectively). W. Mejbaum calls this statement simply dependence.

Hence, correspondence holds between the statement of dependence Z and the statement of dependence Z' if and only if : (1) (condition (a)) in certain circumstances V, dependence Z does not obtain, whereas dependence Z' does obtain in those circumstances, and (2) (condition (b)) when the parameter A assumes the numerical value a, the statement of depeendence Z is synonymous (logically equivalent) with the statement of dependence Z'.

In a special case, the value a of parameter A is its 'limiting' value (the 'zeroing' of parameter A or its inverse), and conditions V involve not assigning to parameter A this 'limiting' value a. Then we are dealing with the very situation distinguished by the conception of correspondence presented by Z. Augustynek. Yet W. Mejbaum's concept also has a strictly generalizing character, since dependence Z' is a generalization of dependence Z across additional circumstances V (i.e., 'non-limiting' circumstances). Analogous remarks can be applied (*mutatis-mutandis*) to both understandings of correspondence. Primary among these is that Z can only be a corrected correspondence translation of some 'original' statement of the relevant dependence, corrected by the fact of its exclusive reference to circumstances other than V. Because it is assumed here that this corrected translation would have to be synonymous (identical) with the original statement of dependence— restricted to non-V circumstances, it is easily seen that what we have here is a conception of correspondence as an 'internal' relation, a conception, what is more, that belongs to the first group and is therefore strictly generalizing.

The most important intuition expressed in the first of our three examples of the traditional conception of correspondence can be put into words more or less as follows : we have an original statement of the dependence of E on A, and then, in the course of the further development of scientific knowledge, we generalize it as a statement of the dependence of E' on A'' in such a way that the denotation of notion A (A') is properly contained in the denotation of notion A''. On the other hand, the next two examples are concerned with an intuition of more or less the following sort : we have an original statement of the dependence of E on A, and then, in the course of the further development of scientific knowledge, we notice that under circumstances V, this dependence does not hold. Fortunately, however, we find a dependence which holds even under circumstances V and which has the important property that its statement yields the original statement of dependence—restricted of course to situations of the non-V type. The fact that this is not actually the original statement corrected, but its corrected correspondence translation (expressing the dependence of E' on A' under non-V conditions) goes unnoticed, but then again there is no need to notice this, if we relate these concepts exclusively to appropriate cases of 'internal' correspondence. For in the context of this correspondence, a correspondence translation may in fact be synonymous with or even identical to the correspondent statement or to a restricted version of that statement, as strictly generalizing concepts of correspondence prefer.

All the examples presented above of concepts of the mentioned kind failed more closely to qualify the type of correspondent statements from the viewpoint of the dichotomy : 'observational' knowledge *vs.* 'theoretical' knowledge. This very qualification, which we will not analyze closely in this chapter (we'll take the matter up in the next chapter), appears, for instance, in the article by S. Amsterdamski 'Spór o koncepcję postępu w nauce' ('The Controversy over the Concept of Progress in Science')[14]. In accordance with an outlook that was quite common in western philosophy until not long ago, he accepts that the correspondent theorems represent 'facts and empirical laws which the old theory explained' and which 'must also find an explanation on the grounds of the new'[15], whereas explanatory premises of the old theory do not themselves have to be correspondent theorems.

Hence, from this point of view, correspondent theorems belong to the phenomenalistic part of every theory. In other words, they are always formulated in 'observational' terms. An accurate intuition is bound up with this outlook, but we will not concern ourselves with it until the next chapter.

Thus, the examples examined so far represent the following variants of the 'internal', strictly generalizing concept of correspondence.

1. The conception that does not make qualification with respect to the dichotomy : 'theoretic' knowledge *vs.* 'observational' knowledge :

a. correspondent statement is synonymous with (identical to) its simple correspondence translation ;

b. correspondent statement, restricted to definite 'limiting' conditions, is synonymous with (identical to) its appropriately corrected correspondence translation ;

c. correspondent statement, restricted to definite, special conditions, is synonymous with (identical to) its appropriately corrected correspondence translation.

2. Conception assuming that correspondent statements have an 'observational' character.

A more detailed illustration of conception 2 is actually extraneous here. For it suffices to analyse the conception which does not qualify with respect to the dichotomy : 'theoretic' knowledge *vs.* 'observational' knowledge in order to reconstruct for oneself the relevant variants of the conception which does take this qualification into account (conception 2). These are also represented by analogous variants (a, b, and c).

The concept of approximately generalizing correspondence embraces those traditional understandings of the relation of interest to us here which have already been attributed to the second group and those which are characterized by the fact that, although they assume the synonymy of correspondence renderings A and A' as well as E and E', they treat the correspondence translation (be it simple or corrected) of the statement of dependence of E on A as only approximately synonymous with the (perhaps appropriately restricted) statement itself and not as its synonym. This correspondence translation, approximately synonymous with the original (possibly restricted) statement, comprises in turn, in the framework of the corresponding system (theory), a special instance of some more general theorem of the system (or theory). It seems that a thorough analysis of individual examples, intended to illustrate this concept of correspondence, would show that we are primarily dealing not with 'internal', generalizing correspondence, but with essentially corrective correspondence. For certainly the idea of this approximate synonymy often expresses in its own way the intuition of the logical incomparability of the original (possibly restricted) correspondent statement and its essentially corrected correspondence translation.

Because the basic difference between concepts of correspondence belonging to the first group and those belonging to the second group stems from the very fact that in the second group, the relation of synonymy between the translated theorem and its (corrected) correspondence translation is replaced (implicitly) by a relation of approximate synonymy, we will confine ourselves here to examining this idea of approximation[16].

Notice that the notion of approximate synonymy is not only intertwined with studies concerning correspondence, but is also assumed or even used explicitly when the problem of explanation is taken up. There is nothing surprising about this, after all, since the structure of 'internally' conceived correspondence is a structure of explanation. The (corrected) correspondence translation, in this context, comprises the explanandum, while its appropriate generalization is the main element of the explanans.

It can be easily observed that the outlook that the (perhaps restricted) correspondent theorem is not synonymous—but only approximately equivalent to—its (corrected) correspondence translation conforms to the idea that 'theoretical' explanation does not in fact apply to the explanandum of which it seems to speak (the original correspondent theorem), but to a somewhat different one—approximately equivalent thereto (the correspondence translation of the correspondent theorem).

This very concept is presented by M. Hesse. She also sets forth the thesis that the explanans in this explanation comprises a kind of metaphorical redescription of the domain of the explanandum[17] (the original explanandum).

In a somewhat broader way, the viewpoint of M. Hesse can be presented as follows. (1) In 'theoretical' explanation, sentence D describing the domain of the explanandum does not in fact result from explanans E. Rather, explanans E yields sentence D', which is in a fairly 'loose' sense equivalent to D. In order for E thereby to be a satisfactory explanans, it is necessary in turn for D' to be a more satisfactory description of the domain of the explanandum than is D. It is necessary, namely, for description D' to render possible the repetition of experimentation with greater precision, to be characterized by a higher degree of compatibility with the entirety of accepted laws than is D, and to make it possible to detect disturbing factors which description D does not take into consideration. (2) The relation of deducibility which is supposed to join D with E fails to come into the picture also because D and E belong to two separate languages : the language of observation and the language of 'theory'. After all, the idea of rules of correspondence[18] which supposedly are to join both these languages is at least dubious. (3) The concept of explanans E as a metaphoric redescription of the domain of the explanandum resolves this situation. From the viewpoint of this redescription, D ceases to be an adequate description of the domain of the explanandum and is replaced by description D', which does result from E. At the same time, we reject the idea of rules of correspondence : '... there are no correspondence rules, and this view is primarily designed to give its own account of the meaning of the language of the explanans. There is *one* language, the observation language, which like all natural languages is continually being extended by metaphoric uses, and hence yields the terminology of the explanans.'[19]

We will not examine M. Hesse's concept[20] in detail here, however worthwhile it might otherwise be to study the ques-

tion of to what degree this concept gives an accurate account
of the situation defined by the appearance of the relation of
essentially corrective correspondence. Presuming that the care-
ful reader will want to resolve this question himself, we will
now devote ourselves exclusively to the relation of approximate synonymy, which in particular is supposed to link the
original (possibly restricted) correspondent theorem with its
(possibly corrected) correspondence translation.

As we see, this is an antisymmetric relation. Description D
(the correspondent theorem, possibly restricted) is less 'satisfactory' than description D' (the correspondence translation,
possibly corrected). This means, according to M. Hesse, that D',
in contrast to D, enables experimentation to be repeated with
greater precision, is characterized by a greater compatibility
with the whole of accepted (as should be surmised—those connected with the corresponding system of knowledge) laws (it
comprises a special case of some of them) and makes it possible
to detect those disturbing factors which were not taken into
account during the formulation of D. Notice with regard to this
last point, that in restricting the generality of D to only defined
conditions, we thereby take at least some of the mentioned
disturbing factors into consideration, although the consideration of disturbances does not have to resolve itself into this
alone.

The problem of reduction, as it is called, from a previously
established theory to a theory that has subsequently arisen is
also intimately linked to the problem of correspondence and
explanation. The concept of 'internal', strictly generalizing
correspondence, which assumes the explanation of the correspondent theory (correspondent statement), perhaps in a restricted form, by the corresponding theory (corresponding statement—a generalization of the correspondence translation),
is matched by the concept of reduction as an operation of the
deductive derivation of one theory (perhaps in its restricted
form) from another theory. Upon examining the problem of
reduction, our attention is repeatedly drawn to the fact that

in numerous cases, reduction does not consist in a simple deductive derivation of one theory from another, but in the deductive derivation from this second theory of only a certain approximation of the first theory. This approximation is 'better' than its equivalent insofar as the latter is subsequently eliminated, replaced by the former. This type of reduction L. Sklar calls a replacement reduction, since in this case, the reduced theory is not derived from the reducing theory, but is replaced by its approximation, which really is derived from the reducing theory.

Notice that Sklar's replacement reduction exhibits many features in common with the 'theoretic' explanation of M. Hesse. L. Sklar, however, as opposed to M. Hesse, does not deem Hempel's popular deductive model of explanation to be useless. In his view, the model possesses a broad range of applications with regard to numerous (it is assumed) cases of reduction of a type other than replacement reduction, with regard to cases, that is, which are characterized by the structure of deductive explanation.

Therefore, to the extent that correspondence comprises a variant of 'theoretic' explanation (M. Hesse) or a variant of replacement reduction (L. Sklar), the correspondence translation of a correspondent statement ought to be regarded as its near synonym and also, respectively, as a consequence of the metaphoric redescription of the statement's domain (H. Hesse) or as an approximation of the statement, resulting from a subsequent system of knowledge (L. Sklar).

It is evident that L. Sklar's concept of replacement reduction has, with reference to the issue of correspondence, a more universal character than does the conception of M. Hesse. The correspondence translation of a correspondent statement is— by virtue of what mental operations it has been achieved notwithstanding (and this may in particular be an approximation of metaphoric redescription)—an approximation of the (possibly restricted) statement. The advocate of M. Hesse's concept will also express agreement with this approximateness. He will agree primarily because this approximation of a (perhaps restricted) correspondent statement (or theory) displays, in comparison with that statement, two characteristic features : (1) it is formulated within a different conceptual apparatus, and (2) it is more 'satisfactory' in M. Hesse's sense already discussed herein.

L. Sklar also specifies more precisely than M. Hesse this more 'satisfactory' character of the approximation of a correspondent statement (theory)—in juxtaposition with the statement (or theory). Among other things, he emphasizes that the approximation (or the theory of which it is a consequence) explains why the correspondent statement (theory) seemed in its own time to be correct—in the context of the experimental procedures in use at the time. More generally, it explains the scientific acceptance of the correspondent statement (and theory)[21].

A fundamental question now arises. Can all these conceptions of correspondence (in a majority of cases, what is more, these are implicitly conceptions of correspondence, since explicitly they speak of 'theoretical' explanation or reduction) which sometimes at least identify the correspondence translation of a correspondent statement (or theory) with its approximation in the sense sketched here validly be treated as conceptions of the 'internal' type ? After all, we are involved here with recognizing features as characteristic of essentially corrective correspondence as the difference in the conceptual apparatuses with the help of which the correspondent statement and its correspondence translation are respectively articulated. (It would seem that this difference might be understood as the difference in relevant methodological-theoretic humanistic coefficients.) Another feature is that the corresponding system of knowledge, containing, in particular, the correspondence translation, allows the range of effective practical applications of the correspondent statement (theory) to be defined and thereby also permits the 'success' (enjoyed in its own time) of that statement (theory) to be explained. Nevertheless, we answer the question affirmatively. Even though traits specific to essentially corrective correspondence are so very recognizable here, the relation is still conceived of as one of the 'internal' type. For though the possibility of establishing logical connections between the correspondent statement (theory) and its correspondence translation is abandoned, according to this position, there remains, as a fundamental connecting bond, a mathematical relationship, consisting in the fact that (1) numerical values which can be computed within the framework of predictions attainable on the basis of the—possibly restricted—correspondent statement match entire intervals to which numerical values belong (or which contain subintervals) that can be computed within the framework of predictions attainable on the basis of the—possibly corrected—correspondence translation, and (2) what is most important here, an algorithm can be discovered which transforms the first method of computation into the second method. Because mathematical 'conversion' is a 'conversion' of the 'internal' type[22] (as is logical 'conversion'), we are still dealing with an 'internal' 'mental' conception of correspondence. It is justified then to number the positions under discussion, treating correspondence as an approximately generalizing relation, among conceptions of the 'internal' type.

2.4. AN EXAMPLE OF ESSENTIALLY CORRECTIVE CORRESPONDENCE. A DEBATE WITH THE VIEWS OF P. K. FEYERABEND

Among arguments directed against Copernican theory, the so--called 'tower argument' figured prominently. Tycho Brahe, among others, considered this argument convincing, but Galileo opposed it. Because Galileo's reasoning in this regard involves an especially simple and therefore didactically advantageous example of essentially corrective correspondence, and because this instance of correspondence and Galileo's argument concerning it provide an opportunity for P. K. Feyerabend to present positions which interpret—differently than the present author—analogous (from the standpoint of interest to us) data from the history of science, we will now examine closely the context of the 'tower argument' mentioned above.

The 'tower argument' may be presented as follows. If the rotational movement of the Earth around its axis were to be accepted, then a physical object dropped from the top of a tower would not fall to its foot. It would not travel a path perpendicular to the tangent of the surface of the Earth at the point located at the tower's foot. It would have to fall a certain distance from that point, since during its fall, the Earth together with the tower would move somewhat in accordance with the rotational motion. Yet this is not what happens. Physical objects dropped from the top of the tower fall precisely to its foot, travelling—as everyone can see perfectly well—a path perpendicular with respect to the mentioned tangent.

How does Galileo counter the above reasoning ?

In the *Dialogue Concerning the Two Chief World Systems, Ptolemaic and Copernican*, this reasoning is represented by Simplicio, who generally appears in the *Dialogue* as a spokesman for common practical wisdom and for the 'metaphysical' (in a positivistic sense) ontology of the Peripatetics, which was specifically related to this common practical wisdom and recorded it scrupulously. He states namely that the correctness of the reasoning under discussion can be shown 'by means of the senses, which assure us that the tower is straight and perpendicular, and which show us that a falling stone goes along grazing it, without deviating a hairsbreadth to one side or the other, and strikes at the foot of the tower exactly under the place from which it was dropped'[23].

This reasoning is countered by Galileo with the following statement of Salviati (who, among the three figures of the *Dialogue*, expresses the outlook of the 'Academic', i.e., of Galileo himself, in the most adequate way): if the earth were, however, to rotate around its axis, then the motion of the stone thrown from the tower 'would have to be a compound of two motions ; the one with which it measures the tower, and the other with which it follows it. From this compounding it would follow that the rock would no longer describe that simple straight perpendicular line, but a slanting one, and perhaps not straight'[24]. For to accept the supposed evidence of our senses is to assume in advance the (rotational) quiescence of the Earth. This evidence in itself, after all, does not imply the (rotational) quiescence of the Earth even as the fact that a 'stone dropped from the top of the mast when the ship was sailing rapidly fell in exactly the same place on the ship to which it fell when the ship was standing still'[25] does not imply that the ship was not moving.

The reason why the characteristics of the given motion cannot be derived directly from the evidence of our sense is, in turn, that this evidence reveals only one aspect of the motion, while another aspect it possesses remains imperceptible to our senses. 'We never see anything but the simple downward (motion), since this other circular one, common to the Earth, the tower and ourselves, remains imperceptible and as if nonexistent. Only that (motion) of the stone not shared by us remains perceptible, and of this our senses show that it is along a straight line always parallel to a tower which is built upright and perpendicular on the surface of the Earth'[26].

Thus, the actual motion, conceived of in the light of both these aspects : the sensorily perceptible and the sensorily imperceptible, is a kind of composite or ensemble of the two. Observationally, we apperceive (in the sense of perceiving cognizantly) this ensemble, this actual motion, to the very extent that we ourselves do not participate on it. For the actual motion is a relative motion—relativized to a frame of reference that wholly fails to participate in it. We could fully perceive this motion, in its complete ensemble, only if we wholly fail to take the observational position of that reference frame. Therefore we ought not to trust indiscriminately every bit of sensory evidence taking from under various (from the viewpoint of our discussion) circumstances. In particular, 'with respect to the

Earth, the tower, and ourselves, all of which keep moving with the diurnal motion along with the stone, the diurnal movement is as if it did not exist; it remains insensible, imperceptible, and without any effect whatever. All that remains observable is the motion which we lack, and that is the grazing drop to the base of the tower.'[27]

It is now easily noticed that for the Peripatetic, representing the viewpoint of common practical wisdom—the action of the letting go of an object such as a stone from the top of a tower—let's denote this action (to return to our previously applied designation) by A—is connected with its effect, E: the falling of the stone at the foot of the tower—in a totally different manner than is action A', manifesting itself in virtually the same way as A, with its respective effect, E' (which also manifests itself in virtually the same way as does effect E) from the point of view represented in the *Dialogue* by Galileo (more precisely speaking—by Salviati). In the first case, namely, this connection involves the translocation of the stone along a straight line, parallel to the tower. In the latter case, however, we are dealing with a composite motion. One of the components is a vertical (downward) motion. The other is the circular motion of inertia[28] (linked with the rotational movement of the Earth). The actual motion, relativized to a frame of reference located beyond the Earth, is the resultant of both these component vectors, and also possesses—as Galileo would have it—a circular character.[29]

Let variable x range across a set of physical objects such as a stone. Then action A can be formulated with the help of the molecular predicate: 'at time t_i, x occupies position P located above the surface of the Earth and at time t_i, x is dropped'. On the other hand, effect E is expressed by the predicate: 'at time t_j (subsequent to t_i), x occupies position R and position R is located on the surface of the Earth on the shortest of line segments located on lines passing through position P'. At the same time, action A' is expressed by the predicate: 'at time t_i—from the viewpoint of reference frame u—x occupies position

P', located above the surface of the Earth and at time t_i, x is dropped', while effect E' is expressed by the predicate : 'at time t_j (subsequent to t_i)—from the viewpoint of reference frame u—x occupies position R' and position R' is located on the surface of the Earth on an arc which passes through P' and the radius of which extends from the median point of the segment : centre of the Earth—position P''[30].

The two pairs of predicates will be abbreviated, respectively, by '$U(x)$' and '$S(x)$' and by '$U'(x)$' and '$S'(x)$'. We can now state that predicates '$U(x)$' and '$U'(x)$' as well as '$S(x)$' and '$S'(x)$' differ from each other even with respect to grammar. For the second elements of each pair are in fact two-argument predicates ; their second arguments are names of specific reference frames (symbolized in the full formulae by 'u'). Predicates '$U'(x)$' and '$S'(x)$'—as they have been formulated above—assume a frame of reference absolutely non-participatory in the rotational motion of the Earth. These have after all been formulated on the basis of that passage in Galileo's discussion[31] which aimed at showing the maximum difference between the relative and absolute motion of a falling object. Nevertheless, these predicates should be formulated more generally. Any frame of reference u can be freely selected. In conjunction therewith, the greater the degree to which a reference frame participates in the rotational motion of the Earth, the more weakly the circular component of the falling object's motion is pronounced. It vanishes completely upon identification of the frame of reference with the initial position P' of the object. Of course, even such generalized predicates '$U'(x)$' and '$S'(x)$'— we will refer to these generalized predicates as '$U''_u(x)$' and '$S''_u(x)$'—will still denote different relations (that is, binary) than do the one-argument predicates '$U(x)$' and '$S(x)$'.

Now the denotations of predicates '$U''_u(x)$' and '$S''_u(x)$' are correspondence renderings of the denotations of predicates '$U(x)$' and '$S(x)$'. For the predicates of the first of these pairs denote sets of situations manifesting the same directly observable (in practice) traits as do the situations belonging to the denotations of predicates '$U(x)$' and '$S(x)$', respectively (i.e., the letting go of a specific object from a certain height and its falling at a specific spot). At the same time, the conditional sentence :'$\bigwedge_{x} [U''_u (x) \to S''_u (x)]$', restricted to cases in which the frame of reference is identical with the initial position P' of the falling object, is the correspondence translation of the conditional sentence:'$\bigwedge_{x} [U(x) \to S(x)]$'. This is clearly an essentially corrected translation, since Galileo's concept of relative motion and the Aristotelian (and thus, the common) absolutist understanding of motion comprise radically different methodological-theoretic coefficients of ideas denoting sets of the relevant situations, which manifest themselves identically on the direct-practical plane. A reflex of this difference, expressed explicitly in the formulation of the predicates '$U(x)$' and '$S(x)$' vs. '$U'(x)$' ('$U''_u (x)$') and '$S'(x)$' ('$S''_u (x)$'), is the grammatical distinction arising among the mentioned predicates. The difference is visible in a most 'spectacular' way in that the actual motion of a falling object takes place, from one point of view, along a straight line, and from the other viewpoint, along an arc.

We will now turn our attention to the Feyerabend analysis of the example of essentially corrective correspondence under discussion. Within his position, expressed in the results of the analysis, I distinguish two component points. The first point— from the stance we assume here—can be accepted, although only after a radical modification. The second point, however, we will confront with a polemic.

Let's begin with the first point. In *Against Method*, P. K. Feyerabend states that the inconsistency of a theory with experimentation can be removed by the very means Galileo employed in connection with the 'tower argument' advanced against the Copernican theory. It might be pointed out here that my catch-phrase 'removing inconsistency' is used somewhat inappropriately in this context. Yet this question is linked with the second of the two points mentioned above and we will therefore return to it later on. In any event, the method used by Galileo, in P. K. Feyerabend's opinion, involves changing the so-called 'natural interpretation' of observational data, in this case—data establishing the vertical motion of a stone falling from a tower.

'Natural interpretations' are not something slapped, as it were, onto observational data : they articulate and organize the data as well. 'There are not two acts—one, noticing a phenomenon : the other, expressing it with the help of the appropriate statement—but only one, viz. saying in a certain observational situation, «the moon is following me», or, «the stone is falling straight down». We may, of course, abstractly subdivide this process into parts, and we may also try to create a situation where statement and phenomenon seem to be psychologically apart and waiting to be related. (This is rather difficult to achieve and is perhaps entirely impossible.) But under normal circumstances such a division does not occur ; describing a familiar situation is, for the speaker, an event in which statement and phenomenon are firmly glued together.

This unity is the result of a process of learning that starts in one's childhood. From our very early days we learn to react to situations with the appropriate responses, linguistic or otherwise. The teaching procedures both shape the «appearance», or «phenomenon» and establish a firm connection with words, so that finally the phenomena seem to speak for themselves without outside help or extraneous knowledge. They are what the associated statements assert them to be. The language they «speak» is, of course, influenced by the beliefs of earlier generations which have been held for so long that they no longer appear as separate principles, but enter the terms of everyday discourse, and, after the prescribed training, seem to emerge from the things themselves.'[32] Assuming, nevertheless, an abstract separation of 'sensations and those «mental operations which follow so closely upon the senses» (Bacon), and which are so firmly conected with their reactions that a separation is difficult to achieve, [...] I shall call them natural interpretations'[33].

Historically, these intuitively sensed natural interpretations 'have been regarded either as a priori presuppositions of science, or else as prejudices which must be removed before any serious examination can begin. The first view is that of Kant, and [. . .] of some contemporary linguistic philosophers. The second view is due to Bacon . . .'[34] On the other hand, 'Galileo is one of those rare thinkers who neither wants forever to retain natural interpretations nor altogether to eliminate them. [. . .] He insists upon a critical discussion to decide which natural interpretations can be kept and which must be replaced'[35].

One might agree with P. K. Feyerabend's general viewpoint on 'natural interpretations' related above as well as with the characterization drawn on its basis of Galileo's dealing with the 'tower argument', but only under the condition that the view (and the characterization) be stripped of its psychological and individualistic character (methodological individualism is of concern here). After such a transformation, of course, these will no longer be Feyerabend's actual outlooks. For the latter, in fact, sketch only a picture of psychological phenomena taking place within individuals assimilating 'natural interpretations' and—sometimes—overcoming them critically (Galileo). Therefore, given the assumption made in the present study, that the discovering of reality is a social process (more precisely, a practical-social process), the claim that these outlooks describe this process cannot be acknowledged[36].

Thus, appropriately, transforming the outlooks of P. K. Feyerabend related above, we will say that the 'natural interpretation' of a situation formulated in a given statement is simply expressed in a relation associating that statement with its literal reference—relativized to a hitherto traditionally accepted methodological-theoretic humanistic coefficient. The sensations and imaginations of actual individuals are irrelevant here. Regardless of what those individuals sense or what they imagine that they see (hear, etc.), statements formulated by them possess—independently of their own will—literal references defined socially (by means of the socially recognized methodological-theoretic humanistic coefficient). The link between a given statement and that which it (literally) asserts (between it and its literal reference) does not have a psychological character, but rather a semantic, that is (in the light of the viewpoint assumed here), a social-subjective one. In order for the reference to be changed, the social acceptance of a change (in science, this change is usually individually proposed) in the domain of the relevant coefficient is indispensable. It was a process of this very kind that was begun in the case of Galileo, who—essentially—proposed a change, later accepted socially, in the literal denotation ('natural interpretation') of the idea of motion and in particular of the predicates we have designated above as '$U(x)$' and '$S(x)$' (the idea of motion is expressed by a conjunction of predicates of this type). This change required, among other things, a transformation of the actually functioning methodological-theoretic humanistic coefficient, which would involve eliminating the belief of the quiescence of the earth in favour of a belief attributing to the Earth at least a rotational motion. This belief, as we know, had previously been advanced by Copernicus.

In summary, the outlooks of P. K. Feyerabend presented so far are acceptable to the extent that they can be acknowledged to fit within the consequences and assumptions of the statement that the literal references (denotations) of predicates '$U(x)$' and '$S(x)$', the Aristotelian idea of motion, underwent transformation in the context of the new methodological-theoretic humanistic coefficient proposed by Galileo. Moreover, one can certainly agree with the opinion of Feyerabend (taken in isolation) that statements encompassing the above-mentioned ideas appropriately re-interpreted (these we will write as '$U_u''(x)$' and '$S_u''(x)$' in order to point out the change from the literal denotations of predicates '$U(x)$' and '$S(x)$') are logically incomparable with the corresponding statements that assume the traditional coefficient.

We will now engage in a polemic with that outlook of P. K.

Feyerabend's which makes up the previously alluded to second point of his position.

Feyerabend states that Galileo, in order to defend the relativistic idea of motion he proposed from the attacks of advocates of Aristotelian physics, abiding by common practical wisdom, resorted to certain 'psychological tricks'. 'These tricks are very successful : they lead him to victory. [...] They obscure the fact that the experience on which Galileo wants to base the Copernican view is nothing but the result of his own fertile imagination, that it has been invented. They obscure this fact by insinuating that the new results which emerge are known and conceded by all, and need only be called to our attention to appear as the most obvious expression of the truth.'[37] Namely, 'Galileo «reminds» us that there are situations in which the non-operative character of shared motion is just as evident and as firmly believed as the idea of the operative character of all motion is in other circumstances. [...] The situations are : events in a boat, in a smoothly moving carriage, and in other systems that contain an observer and permit him to carry out some simple operations'[38]. However, 'the idea of the operative character of all motion [...] arises when a limited object that does not contain too many parts moves in vast and stable surroundings ; for example, when a camel trots through the desert, or when a stone descends from a tower. Now Galileo urges us to «remember» the conditions in which we assert the non-operative character of shared motion in this case also, and to subsume the second case under the first'[39].

The essence, therefore, of these supposed 'psychological tricks' of Galileo seems to boil down to his claiming (as we have already had occasion to observe) that every actual motion has a certain absolute 'inoperative' component and that the intensity of this component is dependent on the degree to which the reference frame of the observer participates in the motion. Absolute 'inoperative' motion would thus be a special case of motion, in the context of which the reference frame is identified with the moving object. It can easily be noticed that this reasoning sets up a relation of correspondence between statements assuming the traditional idea of motion, that is, statements of common practical wisdom, and statements assuming the new relativistic ideas of motion. Due to this, the latter are also substantiated by the former. In the so-called psychological 'tricks' of Galileo, that which for physics (and for all other empirical sciences) comprises a fundamental way to substantiate new theories finds expression : substantiation by demonstrating correspondence with the existing set of research results (with the existing theory).

Galileo, who brought into physics this fundamental means of substantiation which has been so widely used up to our time and which particularly in our times finds such succinct expression, understood it himself, as we see, in the form of correspondence of the 'internal' type : traditional theorems of motion are to constitute a special case of the theorems assuming the relativistic idea of motion, a special case characterized by a specific localization of the reference frame. This is nevertheless an inadequate conceptualization, since this special case of relative motion continues to be relative motion, not identical with traditional absolute motion. The first notion is merely a correspondence rendering of the syntactic category (relativization to reference frame) of the predicate expressing it. We are therefore involved here, as has been already suggested, not with 'nternal' correspondence, but with essentially corrective correspondence. This gives an account of the conditions under which generalizations of common practical wisdom, conceiving of motion in an absolutist way, can be effectively applied in practice, and it thereby allows their social acceptance to be explained. In no way, however, does it lead to further generalizations of these generalizations, but rather to their corrected correspondence translations which are contained in the new theory as consequences of its assumptions. Of course, the corresponding statements are logically incomparable to the correspondent generalizations. Comparability is prevented by the opposition of the respective methodological-theoretic humanistic coefficients.

P. K. Feyerabend, undoubtedly aware of this point, but not taking into consideration the possible appearance of a type of correspondence other than 'internal' correspondence, that is, of essentially corrective correspondence, is glad to interpret Galileo's mental operations, intended to demonstrate a correspondence relation, as a set of 'psychological tricks'. For his concept of 'methodological anarchy', this interpretation is uncommonly handy. Once it is generalized to all cases of essentially corrective correspondence[40], it becomes easy to defend the opinion that the social success of essentially new conceptualizations in a scientific field (and also, in the opinion of Feyerabend, of artistic ideas) is to an important degree determined by the pertinacity of their authors and their ingenious rhetoric. These two factors are sufficient, if in addition the author of a given theoretical-scientific revelation manages to extend it systematically so that it would be possible on the basis of this revelation to articulate new experimental data, forming a new experimental base to confirm it—a base which is not comparable with experimentation 'organized' by and confirming other theories and which therefore is not capable of falsifying those remaining theories.

NOTES

[1] This means that all extra-logical constants appearing in statements (it is assumed that they are sentences of the same language) possess definite literal references (denotations) belonging to the given domain. Here I am using a wider notion of the semantic model of a set of statements : a narrower notion can be obtained by additionally requiring of the model that the statements be true in it.

[2] I analyse this notion of states of affairs more deeply in Z *metodologicznych problemów interpretacji humanistycznej* (*Methodological Problems in Humanistic Interpretation*), Warszawa, 1971, pp. 39–48.

[3] Of course, each individual act of semantic interpretation of a given formalized theory associates this theory with a uniquely selected relational system (a unique domain).

⁴ In fact, the notion of 'standardness' is relativized to a particular group of expressions in the language under interpretation. That is why in rigorously conducted semantic analyses connected with this matter, one should always speak of a standard semantic interpretation with respect to this or that set of expressions (e.g., with respect to the predicate of identity).

⁵ Or else—conversely—naively realist. This possibility does not warrant serious consideration though.

⁶ In spite of this, many Marxist philosophers understand correspondence in an exclusively 'internal' way. This is often true of those who are far from submitting to the influences of positivist naturalism (which conceives of the relationship of a cognizant subject to reality as a psychophysiological relation of that subject to some pre-determined field, not, however, as a relationship of social consciousness to the objective conditions of praxis).

⁷ Generalizing or explanatory correspondence arises, as a rule, only when we conduct some logical operation on system S, for instance, "restriction"—reducing the original generality of its theorems. In this regard, all measures are permissible which can be carried out on the basis of S' within the framework of the methodological-theoretic humanistic coefficient of system S and S'. As it appears, these measures correspond to all the types of reduction (of theoretic systems) distinguished by K. Zamiara in her book *Metodologiczne znaczenie sporu o status poznawczy teorii* (*The Methodological Significance of the Controversy over the Epistemological Status of Theories*), Warszawa, 1974, pp. 16–26—with the exception of explicative reduction.

⁸ If two consecutive systems S and S' are in the relation of correspondence, we will refer to S' as the corresponding system and to S as the correspondent system.

⁹ The conclusion emerges that physics, understood as a science of the objective conditions of praxis carried out on physical objects that continually corrects itself by creating premises which potentially explain the efficacity of praxis based on existing knowledge in the field, is beginning (in this perspective) clearly to take on features generally attributed to the humanities. This matter will not be pursued here, however.

¹⁰ J. With-Hansen, 'Two Methods of Justification and the Correspondence Principle', in K. Ajdukiewicz (ed.), *Foundations of Statements and Decisions*, Warsaw, 1961, p. 85.

¹¹ Z. Augustynek, 'O zasadzie korespondencji' ('On the Principle of Correspondence'), in *Metodologiczne implikacje epistemologii marksistowskiej* (*The Methodological Implications of Marxist Epistemology*), ed. J. Kmita, Warszawa, 1974, p. 243.

[12] W. Mejbaum, 'Prawa i sformułowania' ('Laws and Formulations'), *Studia Filozoficzne* 1962, No. 4.

[13] W. Mejbaum, *Ibid.*, p. 124.

[14] S. Amsterdamski, 'Spór o koncepcję postępu w nauce', *Kwartalnik Historii Nauki i Techniki* 1970, No. 2.

[15] *Ibid.*, p. 50.

[16] Notice that this is a viewpoint precisely corresponding to the instrumentalist concept of the cognitive status of scientific theory. A theory 'says' as much about the world as do the phenomenalistic generalizations resulting from it, whereas the purely theoretic deductive premises for these generalizations constitute only an instrument for arranging and systematizing (in particular, explaining) the former. Cf. in this matter, K. Zamiara, *Metodologiczne znaczenie sporu o status poznawczy teorii* (*The Methodological Significance of the Controversy over the Epistemological Status of Theories*), Warszawa, 1974, pp. 27–68.

[17] Cf. M. Hesse 'The Explanatory Function of Metaphor', in Y. Bar-Hillel (ed.), *Logic, Methodology and Philosophy of Science. Proceedings of the 1964 International Congress*, Amsterdam, 1965, p. 249.

I once engaged in polemics with the thesis of M. Hesse (J. Kmita 'Wyjaśnianie naukowe a metafora' ('Scientific Explanation and Metaphor'), *Studia Filozoficzne* 1967, No. 3), attempting to show that the comparison of explanation with metaphor is misleading in several respects. Nevertheless, certain intuitions linked with this comparison are pertinent. I defined the kind of explanation that conforms to these intuitions as explication (as opposed to reportorial explanation). I would now add that the establishing of the relation of essentially corrective correspondence constitutes a cognitively most significant variant of explication (the 'passage' from one system of knowledge to another system, built with a different conceptual apparatus). In her book, *Metodologiczne znaczenie sporu o status poznawczy teorii* (*The Methodological Significance of the Controversy over the Epistemological Status of Theories*), Warszawa, 1974, pp. 21–26, K. Zamiara constructed— on the basis of this notion of explication—the idea of partially explicative reduction. This would thus be a reduction (of one system of knowledge into another) that assumes essentially corrective correspondence.

[18] Of course, our concern here is with rules of correspondence in the sense of R. Carnap and therefore with postulates of language which assign a so-called 'empirical sense' to theoretic notions. These are only indirectly connected to the idea of correspondence of successive stages of scientific knowledge, which is of interest to us here.

[19] M. Hesse, *Ibid.*, p. 258.

[20] To some extent, this task has been realized by my previously cited

essay 'Wyjaśnianie naukowe a metafora' ('Scientific Explanation and Metaphor'), *Studia Filozoficzne* 1967, No. 3.

[21] Cf. L. Sklar, 'Types of Inter-Theoretic Reduction', *The British Journal for the Philosophy of Science* 1967, Vol. 18, No. 2, p. 112. This point is examined in a yet more elaborate way by K. F. Schaffner in the article 'Approaches to Reduction', *Philosophy of Science*, 1967, Vol. 34, No. 2.

[22] Of course, finding an appropriate mathematical algorithm making it possible to 'convert' from a preceding method of "calculating" to a subsequent one is unusually beneficial in the cognitive and practical exploitation of the relation of essentially corrective correspondence. I only mean that the construction of such a 'conversion' can constitute at most only a symptom that the relation holds, but is not a validated proof of its presence. It is worth noting that certainly the confusion of physical relationships among various parameters with the mathematized representation of these relations—in numerical measures of certain instances of the given parameters—is a source of the identification of mathematical 'conversions' with correspondence 'conversions'.

[23] Galileo Galilei, *Dialogue Concerning the Two Chief World Systems— Ptolemaic and Copernican*, University of California Press, Berkeley and Los Angeles, 1967, p. 139.

[24] *Ibid.*, p. 139.

[25] *Ibid.*, p. 144.

[26] *Ibid.*, p. 163.

[27] *Ibid.*, p. 171. This is Salviati's response to Simplicio's stance that 'good heavens, if it moves slantingly, why do I see it move straight and perpendicular ? This is a bald denial of manifest sense ; and if the senses ought not to be believed, by what other portal shall we enter into philosophizing ?'.

[28] Galileo, as is well known, believed that inertial motion is circular (and not linear).

[29] Galileo presents this motion as follows. Let's consider, radius *AB* of the Earth (*A*—centre of the Earth, *B*—point on its surface, which is located at the foot of the tower). 'I draw the circle *BI* with *A* as a centre and radius *AB*, which reprents the terrestrial globe. Next, prolonging *AB* to *C*, the height of the tower *BC* is drawn ; this, carried by the Earth along the circumference *BI*, marks out with its top the arc *CD*. Now dividing the line *CA* at its midpoint *E*, and taking *E* as a centre and *EC* as radius, the semicircle *CIA* is described, along which I think it very probable that a stone dropped from the top of the tower *C* will move, with a motion composed of the general circular movement and its own straight one. For if equal sections *CF*, *FG*, *GH*, *HL* are marked on the circumference *CD* and straight lines are drawn to the centre *A* from the points *F*, *G*, *H*, and *L*, the parts of these

intercepted between the two circles *CD* and *BI* represent always the same tower *CB*, carried by the Earth's globe toward *DI*. At the points where these lines are cut by the arc of the semicircle *CI* are the places at which the falling stone will be found at various times.' *Ibid.*, p. 165.

30 Cf. preceding footnote. Note also that the formulation of situation E' herein proposed is significantly simplified in relation to the theoretic thought of Galileo, who assumed uniformity of the actual motion of a body in free fall (as opposed to the uniform acceleration of its observable linear component). Yet for the purposes of the present analysis, a fuller consideration of Galileo's principle of relativity is not vital.

31 This passage—or, more precisely, its beginning—was quoted in note 28.

32 P. K. Feyerabend, *Against Method, Outline of an Anarchistic Theory of Knowledge*, London, 1978, p. 72.

33 *Ibid.*, p. 73.

34 *Ibid.*, p. 73.

35 *Ibid.*, p. 74.

36 The question of whether Feyerabend's assertions characterize, with relative adequacy, the individual-psychic phenomena associated with the process under discussion I am leaving aside. I will add, however, that I fully agree with P. K. Feyerabend's opinion of the exceptional status of Galileo as a thinker. He is undoubtedly one of the most outstanding epistemologists in the history of philosophic thought, indisputably surpassing all the philosophers of his day with his shrewdness and the gnosiological significance of the problems he took up. The fact that at the same time he is one of the very greatest exponents of physics ought not to obscure this state of affairs.

37 *Ibid.*, p. 81. In P. K. Feyeraband's view, Galileo bases this method of argumentation on the Platonic theory of anamnesis.

38 *Ibid.*, p. 81–82.

39 *Ibid.*, p. 83–84.

40 In every case that we are dealing, in our terms, with essentially corrective correspondence, P. K. Feyerabend in effect sees a sudden and total 'break' in the developmental continuum. As a consequence, specifically corresponsive usage of correspondent research results qualifies as a 'veritable nightmare' from the 'methodological point of view' (*Ibid.*, p. 62). An example of such a 'nightmare' is to be (among others) the method of determining—on the basis of relativistic mechanics—the orbit of Mercury. It is accepted that the perihelion of the orbit changes about 5600″ every century. This result is accounted for as follows. A change of 5026″ proceeds from the movement of the reference frame. The remaining 575″ result from perturbations within the solar system. And of these 575″, 43″ are calculated using premises specific to relativistic mechanics. The above example is interesting in that it does not

belong to the 'usual' range, so to speak, of ways of using correspondent theorems. These 'usual' ways involving using (as premises) corrected correspondence translations, mistakenly associating them with the correspondent system, whereas in the cited example, equations expressing the appropriate functional dependencies are taken 'word for word' from the correspondent system, in the context of which a situation is described that is not analysed within the corresponding system because of difficulties of a formal-technical nature. Our example involves the situation of central symmetry. Are we therefore dealing, in essence, with a 'methodological scandal', consisting in deduction from a conjunction of premises taken from various, logically incomparable systems ? If we consider the fact that mathematical equations in and of themselves make up an essentially neutral record of functional dependencies among mathematical objects (dependencies assigning definite scalars or vectors to ordered n-tuplets of scalars or vectors), then, first of all, their use as hypothetical, isomorphic equivalents of new dependencies among parameters understood in a new way is not a logical offense. Secondly, the very supposition of this isomorphism, which perhaps will be modified somewhat once a proper (corrected) correspondence translation of the relevant correspondent theorems is obtained, is justified in that a general essentially corrective correspondence between the two systems in question has been demonstrated on the strength of other detailed examples. It suffices to accept the first point in the above reasoning in order not to descry a 'methodological scandal' in the example presented by P. K. Feyerabend.

THE OPPOSITION OF THEORY AND EXPERIENCE

3.1. 'DOGMA OF EMPIRICISM'

As is commonly known, W. V. Quine, in debating with the classical version of logical positivism proceeding from the work of philosophers of the Vienna School, distinguished two fundamental—in his opinion—premises of the epistemological orientation they represented. He endeavoured to show that these premises are dogmas accepted without basis and that therefore the entire system of theses based on them shares in their dubious nature as well. The first of these dogmas assumes that the entirety of our knowledge is divisible into two distinct classes of statements. The first class includes what are known as analytic statements and the second includes synthetic ones. The next dogma, the 'dogma of reductionism' declares that 'to each ... synthetic statement, there is associated a unique range of possible sensory events such that the occurrence of any of them would add to the likelihood of truth of the statement, and that there is associated also another unique range of possible sensory events whose occurrence would detract from that likelihood"[1].

This very 'dogma of reductionism', to the extent that it constitutes a basis for constructing an opposition : theory *vs.* experience, will be the focus of our inquiry in the first section of this essay.

W. V. Quine singles out two separate versions of the reductionism characteristic of two successive developmental phases of the conception of logical empiricism. The earlier as well as stronger version, referred to by Quine as radical reductionism, postulates the translatability (synonymy) of all synthetic statements into (with) 'statements about immediate experience'. These statements about immediate experience are quite simply statements describing individual elements of a set 'of possible sensory events such that the occurrence of any of them would add to the likelihood of truth' of the given synthetic statement that is synonymous with their conjunction (or with their set). In particular, the conjunction may consist of a single component. In turn, the negation of a given synthetic statement is (ought to be) synonymous with a conjunction (set) of other statements, which also describe 'possible sensory events'. The occurrence of these latter is supposed to reduce the likelihood of truth of the given statement. However, a later version (from the time of R. Carnap's *Testability and Meaning*), which is also a weaker, 'attenuated' one, no longer assumes this general synonymy. A looser relation is permitted which can be expressed by means of the so-called dispositional (or—more generally—partial) definitions.

The most representative exponent and, as a matter of fact, the creator of the 'canonical' forms of both versions of the dogma of reductionism was as is commonly known, R. Carnap. The first and strongest version was already introduced by Carnap in *Logische Aufbau der Welt* ; the second 'attenuated' version appeared in *Testability and Meaning*. Quine does not take note however of a third version, which may be called the most attenuated one. Carnap presented it in his essay *The Methodological Character of Theoretical Concepts*. We shall focus here on this final form of the dogma of reductionism, a form which for many philosophers is operative to this very day. We will see shortly that the objections put forth by the author of *Two Dogmas of Empiricism* in principle apply to this third conception of Carnap's as well.

The first version of reductionism can be most concisely described as assuming that every synthetic scientific statement can be translated into a synonymous conjunction of corresponding observational sentences or into a synonymous set of such sentences. (A set enters into consideration when our concern is with a theorem of strict generality, formulated in a language in which there is an infinite class of unit terms forming observational sentences in co-operation with suitable predicates.) The second version, on the other hand, does not treat

this full translatability as a universal feature of synthetic scientific statements, but for at least some statements of this sort, a conditional translatability is posited that is relativized to a specific observational event. The sentence : 'This substance is an acid'—to resort to a standard example—is translatable into the observational sentence : 'This substance turns litmus paper red', under the condition however that an observational event has taken place : the submersion of litmus paper into the substance in question. Thus, the thesis of full translatability of every synthetic scientific statement into an observational sentence, which can also be expressed as a thesis of the translatability of all synthetic scientific statements into an observational language (a language whose only descriptive predicates—predicates other than logical or mathematical ones—are observational), the thesis of the first version of reductionism, gave way to a thesis of full or conditional translatability in the framework of the second version. On the other hand, the third and final version of reductionism accepts—for certain synthetic scientific statements—the possibility of total non-translatability. These statements are not even conditionally translatable into a language of observation. The statements characterized by this non-translatability, in the understanding of this third version, are theoretical statements. The analysis carried out by Carnap of these theoretical statements and especially of their relationship to statements formulated in the language of observation compose an expression of perhaps the clearest traditional understanding—known to this day—of the opposition of theory and experience.

Before this understanding is presented more closely, it is fitting that we point out that a variety of manners of understanding the theoreticity feature of scientific knowledge are commonly operative. In order more clearly to grasp the nature of this feature—when it appears as the opposite of the feature of observability (of direct availability through experience), it is beneficial to bear in mind that theoretical knowledge (theory) is also commonly understood : (1) as the opposite of practical activity ('theory' in this context is simply a name for knowledge forming predictive premises for activity), and (2) as the opposite of factographical knowledge of the idiographic type ('theory' in this context is a name for konwledge which is a system of statements of strict generality). We are interested here, on the other hand, in the theoreticity feature as a property specific to knowledge which cannot be formulated in an observational language and therefore to knowledge which requires for its substantiation other premises in the form (it is traditionally believed) of statements belonging to that language. It is this intuition of the variant of the theoreticity feature relevant to our discussion which is proper to all continuations of the idea of traditional empiricism. Carnap's concept belongs to the group of best known continuations of that idea. However the understanding of the opposition : theory *vs.* experience which will be presented later on in this essay actually maintains only one trait of the intuition in question : the thought that theoretical statements require substantiation effected with the help of premises from beyond the theoretical sphere. Yet in the conceptualization to be proposed herein, these premises are not characterized by their inclusion in a set of beliefs noting certain observable facts as understood within traditional empiricism (and hence, to a set of beliefs which can be recorded in the logical positivist language of observation).

Let us return for now to the conception R. Carnap presented in his essay *The Methodological Character of Theoretical Concepts.*

We observe first of all that in this essay, alongside the idea of an observational language previously used by Carnap in *Testability and Meaning,* a new idea of an extended observational language is employed. Whereas in the observational language, as it had previously been understood by Carnap, only those descriptive predicates could appear which were either (1) primitive and such that the sum of the denotations of any one of these predicates and its negation would be identical with the universe of (observable) objects or events under examination, or (2) fully definable by means of primitive predicates, in the extended observational language, descriptive predicates may also appear which are defined only partially, primarily dispositionally, by means of descriptive predicates of the strict observational language (as we will hereafter call the non-extended observational language of Carnap). It is easily noted that predicates only partially defined by means of predicates of the strict observational language no longer possess the recently stipulated property of the latter. That is, it is not true that the sum of the denotation of each such predicate and the denotation of its negation is identical with the universe. For if a predicate 'Q' is defined[2] by means of two conditional sentences, the first of which asserts that whatever is P is also Q, while the second of which asserts that whatever is R is not Q (where predicates 'P' and 'R' are from the strict observational language), then— as long as the sum of the denotations of P and R does not comprise the universe—beyond the range of the sum of the denotations of Q and non-Q there remains a subset of the universe, of greater or lesser size, constituting the field of characteristic vagueness of predicate Q. The sense of this predicate was established in such a way that neither Q nor its negation refers to the elements of that field. Predicate Q is thus only partially semantically interpretable,[3] whereas predicates of the strict observational language are completely interpretable.

The complete semantic interpretation of primitive predicates of the strict observational language has a 'natural' character, as it were. It is not established via agreement or convention. In this respect, it can be defined as a 'natural' empirical interpretation. The point is that the relation of the fact of predication by each of these predicates of a given object or observable event with that object or event can be expressed by means of a synthetic statement which would specify the determinant of this predication as the appearance of certain physical features of the given object or event (the 'objective', physicist[4] interpretation) or also as the appearance of a certain psycho-physiological state in the observer (the 'subjective', psychological interpretation). We learn the usage of these predicates in an ostensive way as well : individuals who have submitted in a more advanced fashion to the mechanism determining the predication of primitive observational predicates point out concrete examples of the denotations of these predicates (or their negations) to individuals who are less advanced in this respect. Thanks to this and to the process of inductive generalization (a process which emerges spontaneously, since it constitutes the effect of the 'operation' of appropriate psychological regularities), we gradually assimilate the 'correct' empirical interpretation of the given predicates.

The factor of convention arises only in connection with predicates defined wholly or partially by means of primitive observational predicates. Nevertheless, even these possess—due precisely to their whole or partial capacity of definition in terms of those observational primitives—a 'non-conventional' link to (observable) physical or psychic reality, respectively. For just this reason, we are free to interpret knowledge formulated in the extended observational language realistically as a set of statements truly or falsely describing that reality.

The outline we have sketched of this knowledge, which of course also contains statements of strict generality as well as existentially quantified statements, an outline consisting in the realistic interpretability of that knowledge, comprises one of the most important features distinguishing it from theoretical knowledge, i.e., from theory. A second such feature, which moreover, in Carnap's conception (as we have seen), substantiates the emergence of the former feature, is the 'naturalness' of the empirical semantic interpretation of this knowledge. A scientific theory possesses neither of these features. For a theory arises via construction as a so-called 'pure calculus'. It is an empirically uninterpreted axiomatic system that is based on appropriate postulates. The arbitrariness of the system is limited by the consideration of two points : (1) the system must be suitable for 'calculative' manipulations (these involve, for example, the so-called—and diversely conceptualized—requirement of simplicity and the even less clear requirement of convenience, etc.) and (2) it must be true, upon attaching to the system an appropriate additional system of postulates called rules of correspondence[5], that the resulting conjunction allows specific sentences from the extended observational language to be deduced. These, in particular, include generalizations constituting 'empirical laws'. (The quotation marks, of course, are not Carnap's.) This very fact of the conventional 'gluing' together of knowledge from the extended observational language and theory renders a realist understanding of theoretic statements impossible. For rules of correspondence differ from whole or partial definitions used in the extended observational language in that they are a 'mobile' implement, as it were, used en bloc, for linking theory with experience (i.e., with knowledge formulated in the extended observational language). This yields a third, essential difference that holds between theory and observational knowledge : the former can be revoked, whereas the latter—in principle—is not subject to repudiation. This revocability, of course, does not indicate a permanent variability of the scientific-theoretical world image, since theory, as we have seen, ought not to be interpreted realistically and should not be associated seriously (that is, other than just heuristically) to any image of the real world. It is only an instrument (the instrumentalist interpretation of the cognitive status of scientific theory) of deductive arrangement—in a framework of explanation and prediction—of realistically conceptualized knowledge expressed in the extended observational language.

Taking note of this third opposition in the features of theory and experience leads Carnap to a certain change in the methodological recommendations which he formerly directed to psychologists. Namely, at the time of *Testability and Meaning*, when he still believed that all descriptive scientific predicates are (ought to be) either primitive predicates or predicates wholly or partially definable by means of primitives, he required of psychology that newly introduced terms which are not of

a primitive observational nature should be defined dispositionally (primarily because the possibility of total definability
in advance did not seem realistic). Of course, these dispositionally defined predicates would be stated—upon manifestation of
the appropriate observational data—in an irrevocable manner.
Yet this recommendation (or assertion even) is contradicted by
the fact that psychologists, in spite of the appearance of specified observational data which would seem by definition to
determine the affirmative (or negative) predication of some of
their technical terms, do not predicate accordingly, sometimes
even employing the opposite predication. Carnap's explanation
for this fact, from the perspective of *The Methodological Character of Theoretical Concepts*, is that these are not dispositionally defined terms, but theoretical ones. The statements predi-
cating them (or their negations) belong to the theoretical lan-
gauge and are related to statements of the extended observational
language through 'mobile', as it were, systems of rules of correspondence. The replacement of one system of this sort by
another system may cause the annulment of the corresponding
theoretical statements even though the observational statements have not undergone change. Thus, bearing in mind the
fact described above, Carnap is now capable of accounting for
it with the help of his new concept of theory. He now recom-
mends not that psychologists define their necessary technical
terms dispositionally, but rather that they introduce them as
purely 'calculative' terms characterized (syntactically) by postulates, and that they link definite theoretical statements containing them with statements of the extended observational
language by means of the appropriate rules of correspondence.

The above recommendation, as we have seen, follows from
the fact that in psychology, numerous cases can be attested
in which certain terms are used in such a manner that for the
purpose of their predication, the appearance of the appropriate
observational data is not absolutely decisive, but '... if a scien-
tist has decided to use a certain term M in such a way, that for
certain sentences about M, any possible observational results

can never be absolutely conclusive evidence but at best evidence yielding a high probability, then the appropriate place for M in a dual-language system like our system L_0—L_T is in L_T rather than in L_0 or L'_0'.[6] Regardless of the fact that the recommendation of interest to us here is simply meant to sanction a situation appearing in psychological practice (as well as in the practice of all empirical sciences), it rests simultaneously on the belief that '... more comprehensive use of this method will lead in time to theories much more powerful for explanation and prediction than those theories which keep close to observables ... The germs of this development can sometimes be found in much earlier periods and even, it seems to me, in some prescientific concepts of everyday language, both in the physical and psychological field'[7].

Although Quine's argumentation against 'the dogma of reductionism' directly concerns that version which assumes that the whole of scientific empirical knowledge can be formulated in the extended observational language, it can nevertheless be also applied, as has already been indicated, to the final, most liberal version, which we have just discussed. Incidentally, even if that possibility did not enter into the picture, this argumentation still would undermine the concept of the observational-theoretical 'bi-linguality' of scientific knowledge by placing in question the existence of an observational component in the form this conception postulates. We will now consider what W. V. Quine would find objectionable in the final version.

As has been said, three oppositions set knowledge formulated in the extended observational language apart from theoretical knowledge (formulated in the theoretical language). These are : (1) the realist status of the former versus the instrumentalist status of the latter, (2) the steadfast and unambiguous nature of the empirical interpretation of descriptive predicates of the (extended) observational language, built on the 'natural' empirical interpretation of primitive observational predicates, versus the conventional and variable nature of the correspondence rules joining theoretical knowledge with knowledge formulated in the (extended) observational language, and (3) the irrevocability (in principle) of statements formulated in this language versus the revocability of theoretical knowledge.

It is primarily opposition (2) that is attacked by the arguments of the author of *Two Dogmas of Empiricism*. This opposition, as one can easily observe, is a point of departure for the reconstruction of the remaining two oppositions. Let us recall the most fundamental arguments.

First of all, there is no such thing as the 'natural' empirical interpretation of primitive observational predicates. In reference to the use of any given language, that which may be observed by purely empirical means is the manner in which the so-called occasion sentences are used. These are of the type 'Rabbit !' 'Red !', 'Water !'. These sentences, moreover, are not the group to which scientific statements belong, since the latter are always standing sentences. Yet upon observing how these occasion sentences are used, without making any assumptions concerning the language under investigation (a model example of this would be the behaviour of a linguist who wants to learn a language totally unknown to him, that is used by a tribe among whom he is staying by himself with no opportunity for verbal communication with the natives), we can only establish that whenever a physical context of a certain type arises, to the investigator's interrogative use of a given occasion sentence, let's say 'Gagavai ?', representatives of the language being researched react with a gesture of affirmation, whereas whenever a context of some other type arises, a gesture of denial occurs in response to the given question. In this way, we can determine the 'stimulus meaning' of a given occasion sentence ; this may consist of a physical situation, characterized, shall we say, by the presence of a rabbit. Nevertheless, we cannot establish the literal reference of the sentence in an empirically determined manner, which would permit, in turn, the translation of the sentence into the proper standing sentence

of the investigator's language. After all, we do not have decisive data at our disposal. Does, let's say, the occasion sentence
'Gagavai !' refer to the rabbit or to one of its parts, to its temporal stage, to 'rabbithood', or to some kind of immediate data ?
As linguists, we must resolve this question one way or another,
and we resolve it in an empirically undetermined way, following simply the directive that our hypotheses about the literal
references of individual occasion sentences ('analytic hypotheses',
as they are called) should be consistent with one another
and with the stimulus meaning of the sentences themselves
and, finally, that the image we obtain of the language being
researched should be as simple as possible and efficient with
respect to our translating activity, in which we will be assigning equivalents from our own language to the terms and standing sentences of the language under study.

Thus, the denotations of the so-called primitive observational
predicates are not especially 'naturally' empirically determined.
Whether we assign to them one or another set of defined observational relations (including unary ones) as their references
is a question of our own decision, to a significant extent an
arbitrary one. Therefore, there are in fact no primitive observational predicates (in Carnap's sense). There are none, not so
much because we are unable to detect them in purely empirical research into language as because we assimilite our own
mother tongue in precisely the same way as we investigate
a foreign language totally unknown to us : the terms of our
own language obtain definite references (similarly we assign
definite references to the terms of a language under study) only
when we relativize these references to a complete, consistent
system of ontological-semantic decisions, decisions which are
not dictated to us in some "natural" unequivocal way by observational data.

Let us reconstruct the setting in which our native language
is acquired '... with all its predicates and auxiliary devices.
This vocabulary includes «rabbit», «rabbit part», «rabbit stage»,

«formula», «number», «ox», «cattle» ; also the two-place predicates of identity and difference, and other logical particles.
In these terms we can say in so many words that this is a formula and that a number, this is a rabbit and that a rabbit part,
this and that the same rabbit, and this and that different parts.
In just those words. This network of terms and predicates and
auxiliary devices is, in relativity jargon, our frame of reference, or coordinate system. Relative to it we can and do talk
meaningfully and distinctively of rabbits and parts, numbers
and formulas. Next .. we contemplate alternative denotations
for our familiar terms. We begin to appreciate that a grand and
ingenious permutation of these denotations, along with com-
pensatory adjustments in the interpretations of the auxiliary
particles, might still accommodate all existing speech dispositions. This was the inscrutability of reference, applied to ourselves ; and it made nonsense of reference. Fair enough ; reference is nonsense except relative to a coordinate system. In
this principle of relativity lies the resolution of our quandary"[8].

Secondly, the statement that other than primitive observational predicates, the only descriptive predicates that appear
in the extended observational language are synonymous with
appropriate molecular configurations of primitive predicates or
are at least partially semantically determined by such appropriate molecular configurations is untenable since not only can
the ideas of meaning and synonymy not be characterized in
a satisfactory way, but no such characterization at all is possible. It is impossible precisely because the relation of synonymy, however it is conceived of, does not assume any relativization to the entirety of the 'coordinate system'. Incidentally, Quine's critique of the notion of synonymy is inseparable
from his critique of the intensional notions (proposition, property in the intensional sense, etc.) which define it or are definable by it. As a result, it is inseparable from the critique of
the second logical positivist dogma (alongside the 'dogma of
reductionism') : the conviction that the totality of scientific

statements is divisible into the class of analytic sentences and the class of synthetic sentences. We will not delve into this question here, however.

In any event, in the viewpoint of the author of *Ontological Relativity*, the opposition : the 'naturally', empirically interpreted knowledge from the realm of the extended observational language, as knowledge which is not conventional in principle, versus theoretical knowledge, which is conventional in principle, is untenable.

Consequently, neither can the following opposition be maintained : the irrevocability of knowledge from the realm of the observational language versus the revocability of the theoretical knowledge. All elements of our entire scientific knowledge about the world can be countermanded. They can be countermanded because they form a whole, a whole that is bound, among other things, to the system of our ontological-semantic decisions. It is this very whole, moreover, which we confront with the data of observation, and not individual sentences '. . . total science is like a field of force whose boundary conditions are experience. A conflict with experience at the periphery occasions readjustments in the interior of the field . . . The total field is so underdetermined by its boundary conditions, experience, that there is much latitude of choice as to what statements to reevaluate in the light of any single contrary experience.'[9]

A consequence of this position, in turn, is the outlook expressed in what is known as the Duhem–Quine thesis : every hypothesis falsified by appropriate observational statements can be preserved through the assumption of certain additional premises, in particular through the rejection of those observational statements, though this measure is often—for technical reasons—unprofitable. (We will examine this thesis closely in the next chapter.) Therefore, '. . . we can imagine recalcitrant experiences to which we would surely be inclined to accommodate our system by reevaluating just the statement that there are brick houses on Elm Street, together with related statements on the same topic. We can imagine other recalcitrant experiences to which we would be inclined to accommodate our system by reevaluating just the statement that there are no centaurs, along with kindred statements. A recalcitrant experience can, I have urged, be accommodated by any of various alternative reevaluations in various alternative quarters of the total system; but, in the case which we are now imagining, our natural tendency to disturb the total system as little as possible would lead us to focus our revisions upon these specific statements concerning brick houses or centaurs. These statements are felt, therefore, to have a sharper empirical reference than highly theoretical statements of physics or logic or ontology'[10].

The distinction between statements based 'naturally' on experience and theoretical statements (in Carnap's sense) falls through in its turn. How does the matter of this opposition unfold ? As we recall, this opposition assumes the realist status of statements from the domain of the restricted observational language versus the instrumental status of theoretical statements. Thus, the maintenance of this opposition is also senseless from Quine's point of view. Because there are no statements which have references assigned to them unconditionally, independently of the conventional 'coordinate system', none of them is privileged with the opportunity for adequate representation of the corresponding real state of affairs. An instrumental status, that is, characterizes all scientific formulations : those which Carnap would include within the scope of the extended observational language as well as those which in his view are of a theoretical nature. 'As an empiricist', writes Quine, 'I continue to think of the conceptual scheme of science as a tool, ultimately, for predicting future experience in the light of past experience. Physical objects are conceptually imported into the situation as convenient intermediaries—not by definition in terms of experience, but simply as irreducible posits comparable, epistemologically, to the gods of Homer. For my part I do, qua lay physicist, believe in physical objects and not in Homer's gods; and I consider it a scientific error to believe otherwise. But in point of epistemological footing the physical objects and the gods differ only in degree and not in kind. Both sorts of entities enter our conception only as cultural posits. The myth of physical objects is epistemologically superior to most in that it has proved more efficacious than other myths as a device for working a manageable structure into the flux of experience"[11].

Hence, from the point of view of the author of *Ontological Relativity* there is, in the final analysis, no basis whatsoever for distinguishing phenomenalistic knowledge from theoretical knowledge, and there is therefore no possibility of epistemological definition of the notion of theory. Although I will strive later on to reinstate Carnap's opposition in some fashion, it should still not be inferred from this that I lean in favor of Carnap's position. Both that position and the ontological relativism of Quine represent internally inconsistent versions of the same general orientation : traditional empiricism. The orientation assumes the point of view of ahistoricism and of methodological individualism in the process of coming to know reality and, in the final account, is idealistic (since it treats the development of knowledge about the world as the effect of the appearance of specific sensations, mental operations and methodological ideas in the minds (in the organisms) of individual cognizant subjects). The antagonism of these two versions of traditional empiricism is an expression of its internal difficulties and, most of all, the consequence of their single-track consideration of only certain aspects as well as only certain periods in the process of knowledge development. In the formulation of some kind of equivalent (a correspondence rendering, it will be maintained) to Carnap's opposition, I will attempt to show which elements of the two conceptions of interest to us here deserve to be taken into account.

3.2. PERFORMED ACTION AS THE ESSENTIALLY CORRECTED CORRESPONDENCE RENDERING OF UNDERTAKEN ACTION

We will now examine the contrast which, as we shall soon see, constitutes the point of departure for the varying interpretations of the opposition of theory and experience presented in the preceding section. The contrast of interest here is one to which I have referred in various works and which I most extensively treated in the paper *Humanist Interpretation and Functional Explanation*[12] : action undertaken vs. action performed. In its form of that time, the contrast can be presented as follows[13].

We begin with a suitable example. A foreigner undertakes the action of uttering a certain expression in Polish. The sense (purpose) of this expression is to communicate a given state of affairs, and yet it is spoken in a deformed fashion by contrast to the pattern designated by phonological rules that help to make up the linguistic competence defining Polish. Notice that in a situation of this sort, if the deformation is not too extensive, the listener usually surmises what pronunciation 'standard' comes into consideration and classifies under that heading the utterance he has just heard, simultaneously assigning to it the same meaning that is matched to the appropriate standard by rules of semantics. Thus our listener jointly conducts two mental operations : the first operation involves recovering the 'correct' equivalent of the expression he has heard, and the second operation involves a procedure analogous to the one for every 'correct' expression. Parenthetically speaking, expressions uttered by speakers operating in their own native language also seldom happen to be ideally 'correct' (if it ever happens at all), even if we discount speech defects (such as stuttering). As a result, the first mental operation is involved in these cases as well, to a greater or lesser degree of salience.

Yet what does this second operation consist in ? It is, in my conception, idealizing humanist interpretation. (In the event that we are concerned uniquely with this operation, it will normally be called humanist interpretation for short.) It constitutes a certain kind of explanation based on an assumption of rationality that makes the following assertion (only so-called conditions of certainty [14] are taken into consideration here).

If X at time t is to undertake one of the complementary and mutually exclusive—to his knowledge—actions $A_1, \ldots, A_n$, unambiguously associated—to his knowledge—with results $S_1, \ldots, S_m$, ordered in turn—according to his norms—by an appropriate relation of preference, then X at time t will undertake action A_i ($i = 1, \ldots, n$, or A_i is the objective equivalent of the logical disjunction of members of a proper subset of the set $(A_1, \ldots, A_n)$, when all elements of this subset correspond to the same result) associated with the result of maximum preference.

In the context of the variant of explanation now under discussion, the explanandum—as we see—comprises the fact of the undertaking of a given action ; the explanans, on the other hand, contains in addition to the assumption of rationality : (1) a description of the actions $A_1, \ldots, A_n$ relevant in a given concrete instance, as they present themselves from the point of view of the subject's knowledge, (2) a description of the association of actions $A_1, \ldots, A_n$ with results $S_1, \ldots, S_m$, as the association appears from the viewpoint of the subject's knowledge, and (3) a description of the preference relation as it is designated by the norms of the subject and which in particular singles out the preferred result. This is the sense (goal)[15] of the undertaken action.

We note that as long as the knowledge and norms attributed in the act of our explanation to the subject of the undertaken action make up two respective systems of judgments that are wholly consistent and do not contain (deductive) 'gaps', then this attribution has an idealizing nature, since each concrete individual can consciously accept such systems or even simply recognize them only to a certain, imperfect extent. The explanatory premises therefore have an idealizing nature. This is because the normative beliefs and knowledge attributed to the subject constitute two respective, historically current spheres of social consciousness, a consciousness which is always assimilated by specific individuals in a fragmentary and deformed (from various perspectives) way.

Because humanist interpretation is also the name given to the procedure composed of both the mental operations of interest to us here, we will therefore refer to the second operation, which has just been discussed, by the name of idealizing humanist interpretation as was previously mentioned, in order to avoid misunderstandings (when they might otherwise arise). Hence, as we see, idealizing humanist interpretation is a component of another variant of humanist interpretation.

What is the nature of this other variant ? Let's first of all introduce a new terminological convention. An action, the undertaking of which is explained in the framework of idealizing humanist interpretation, will regularly be called an undertaken action. This is a conceptualization of the action from the viewpoint of the socially accepted knowledge attributed to its subject, in other words, an action (a class of actions) constituting the literal reference of the respective concept in the light of the methodological-theoretic humanistic coefficient represented by the knowledge attributed to the subject. At the same time, we will also speak of the intended sense (of an undertaken action). It is easily seen that the example of the first component operation presented above, involving (in the example) the 'subsuming' of an utterance spoken by a foreigner under an ideally 'correct' form, can be generalized in a simple fashion : it is an operation consisting in the 'subsuming' of an actually performed action under the idealized (socially binding) form of an undertaken action.

Notice that in mentally equating the performed action (this is the name we will give it to distinguish it from the undertaken action) with the undertaken action, we conceive of the actual consciousness of the subject of the performed action in a modifying (and idealizing) way. Nevertheless, we do not lose sight of this consciousness when, having established the sense of an action, we wish to explain the fact of just this (and no other) performance of the action. In conjunction therewith, we attribute responsibility for the difference between the undertaken action and the performed action to the difference arising between socially binding knowledge and norms and this concrete variant of them, which is in fact represented by the subject of the performed action. In this context, the primary role is played by the difference in the sphere of knowledge. For it is this which determines the variant articulation of an action and of its sense and—above all else—the variant articulation of the relation between the two, whereas norms only define the preferential ordering of states of affairs articulated in one way or another. This can be seen at least in that identity of the senses of the undertaken and the performed actions is always assumed (even though this is not quite precise).

It might seem that sometimes the 'blame' for 'bad performance' can also be laid on what one might call objective circumstances, such as, for instance, the fact that the subject of a spoken utterance, with a fully adequate knowledge of socially binding phonological rules, stutters anyhow, this constituting "bad performance". This case—in light of the assumptions we have made here—is somewhat more complicated. Yet to illustrate this, it is necessary first to broaden the conceptualization that we have been presenting in an essential way.

Namely, our conceptualization has until now paid attention to only one possible situation. The example of a person speaking in a foreign language illustrates it in an especially poignant fashion. The situation arises when the sole point of reference for the interpreter is a definite system of socially accepted judgments, constituting his idealization of the beliefs actually held by the subject of the action undergoing interpretation. In practice, a very frequent cause of this situation is that the interpreter himself is an advocate of that system of judgments and he is in a position to understand it precisely as an idealization of the set of beliefs of the subject of the action. Alongside this scenario, there is also another possibility, in which the interpreter at the same time considers a given action and its results in the context of a second system of judgments, which may also be socially accepted or simply individually accepted (even if by some group of persons), and in which the first system is neither an idealization of the second nor necessarily at all accepted by the interpreter.

In this case, on the basis of the second system of judgments, the interpreter will speak of that understanding of the interpreted action which is dictated by this second system as a performed action. A performed action understood in this way, moreover, may wholly fail to constitute an object of humanist interpretation, for instance, when we are exclusively interested in the obejctive status of an action, regardless of how it is understood by anyone's consciousness.

When, for instance, Karl Marx states that people, '... by equating their different products to each other in exchange as values ; equate their different kinds of labour as human labour. They do this without being aware of it"[16]—it is clear that he does not have humanist interpretation in mind, nor the establishment of the sense, but rather a characterization of the action of exchange of goods by pointing out its objective effects and its (what is more) objective function from a point of view wholly different from the system of beliefs held by the subjects of the action. The humanist interpretation of a performed action in the understanding we have sketched here comes into play, on the other hand, only when that second system of judgments is attributed to the subject of the action under discussion. Moreover, if this is done in an historically groundless way, the given interpretation ought to be qualified as an adaptation, as opposed to historically substantive interpretation.

Thus, in summary, a statement of the difference between an undertaken action and a performed action may be derived from two separate sources, representing two different conceptions

of the performed action : either (1) that the subject of the performed action operates with a knowledge that is a definite
deformation of a current, socially functioning knowledge, which
constitutes an idealization of the subject's knowledge and defines the undertaken action, or (2) that the performed action
is a somehow competitive—in relationship to the undertaken
action—conception of the 'same' action. This conception, to the
extent that it is connected with humanist interpretation, is
linked by the interpreter to the viewpoint of the subject of the
action. The subject, moreover, may be the interpreter himself
(autointerpretation), as long as he is always an adherent of
this conception. Incidentally, the interpreter, as an adherent of
knowledge fundamentally in opposition to the socially received
knowledge that is current with respect to the performed
action, does not have to be an innovator himself at all. Nor
does he necessarily risk anything by his support for innovation.
The innovative point of view may well have become a part of
social consciousness in the meantime.

We questioned somewhat earlier the opinion that sometimes
the 'bad performance' of an undertaken action is not caused by
an inadequate acquaintance with the context of knowledge that
qualifies the action, but by certain objective circumstances.
(For instance, a stuttering person 'badly performs' particular
linguistic actions even though he is adequately acquainted with
the socially binding beliefs expressed in phonological rules).
In essence, the phrase 'bad performance' is somewhat misleading here. It assumes, after all, a recognition of the relevant
standard or, more precisely, a recognition of the current,
socially functioning knowledge that designates that standard.
At the same time, it challenges that knowledge as ineffective
in the light of the given objective circumstances. Either the
knowledge takes the given objective circumstances into reckoning and is effective in their sphere, while the subject performing the action does not take that knowledge sufficiently into
regard (the case designated above as situation (1)), and his
action is therefore 'badly performed', or else a new knowledge

that is just emerging or has already been established (at least on an individual base), competing with the preceding system, takes these circumstances into account, unlike the old system. This creates the context that defines the performed action as 'badly performed' from the new standpoint (the case designated above as situation (2)), and it does so precisely because (and not in spite of the fact that) the subject abides by the old point of view.

We shall shortly focus on a particular instance of situation (2): the knowledge qualifying a performed action stands in a relation of essentially corrective correspondence to the knowledge qualifying an undertaken action. Yet before we examine this case more closely, we will conclude the current phase of our study with a general description of that variant of humanist interpretation which answers the question of why it is that a given performed action took place and no other and provides an answer that is composed in part by idealizing humanist interpretation. This variant will be called concretized humanist interpretation to distinguish it from the idealizing variant. Concretized humanist interpretation can be presented as an instance of explanation that does not appeal to the assumption of rationality alone, but to the conjunction of that assumption and a statement, which can be called the condition of effectiveness. The condition of effectiveness comes in a variety of forms, depending on the point of reference established by the interpreter. Hence, when the point of reference is the knowledge defining the undertaken action, the condition of effectiveness has the form of a conditional sentence of the type : if the knowledge of the subject of the performed action differs in such-and-such a respect from the (idealizing) knowledge of the subject of the undertaken action, the performed action that takes place will differ from the undertaken action in such-and--such a way. In this situation, the concretized humanist interpretation takes the form of more or less the following reasoning : if X were to accept such-and-such knowledge, then an action undertaken by X would be of such-and-such a kind, but there is a discrepancy, insofar as that knowledge is an idealization of X's knowledge and due to this, an action performed by X takes on such-and-such a shape.

The condition of effectiveness is a statement of another sort in the case where the interpreter operates at the same time from a second point of reference, represented by the subject of the performed action and standing in opposition to the knowledge defining the undertaken action. The latter constitutes a 'negative' (in a certain sense) point of reference for him. The statement is then a conditional sentence of the type : if the knowledge defining the undertaken action differs so radically from the knowledge of the subject of the performed action that the latter is not even an idealization of the former (and in particular, is logically incomparable to it), then a performed action will occur defined in such-and-such a way by the knowledge of the subject of that action. In this situation, the concretized humanist interpretation assumes the form of more or less the following reasoning : if X were characterized by such-and-such knowledge, then an action undertaken by X would be such-and--such, but there is a discrepancy insofar as that knowledge is not even an idealization of X's knowledge and due to this, X performed such-and-such an action (defined in terms of X's own knowledge).

We will name these two variants of concretized humanist interpretation the individualizing concretized and the correctively concretized, respectively. We can now characterize them somewhat more exactly, in the meantime summarizing all the remarks made on this subject until now.

1. Individualizing concretized humanist interpretation

(a) explains the occurrence of a definite (performed) action given the assumption that

(b) the knowledge of the subject of the action differs from the contemporary, socially recognized knowledge in such a way that the latter, in any event, is an idealization of the former ; and

(c) the interpreter characterizes the performed action in

terms of the differences separating it from the undertaken action, which is defined by socially recognized knowledge; and finally

(d) explanation is achieved through deduction of the fact of occurrence of the (performed) action from the counterfactual application of the assumption of rationality, coupled with the condition of effectiveness, which predicts in what respect the action which has occurred (the performed action) will differ from the action which would occur (from the undertaken action) if the subject were operating with the socially recognized knowledge, since there is an applicable discrepancy between that knowledge and the subject's own knowledge.

2. Correctively concretized humanistic interpretation

(a) explains the occurrence of a definite (performed) action given the assumption that

(b) the knowledge of the subject of the action differs from the contemporary, socially recognized knowledge in such a way that the latter is not even an idealization of the former (in particular, they are logically incomparable); and

(c) the interpreter characterizes the performed action exclusively in terms of the subject's knowledge; finally

(d) explanation is achieved through deduction of the fact of occurrence of the (performed) action from the counterfactual application of the assumption of rationality, coupled with the condition of effectiveness, which tells what (performed) action will in fact occur, if the subject possesses knowledge fundamentally differing in such-and-such a respect from the socially recognized knowledge of his times.

It might seem that correctively concretized humanist interpretation is actually a superfluous procedure. Since the subject of the action, as is assumed, is not represented—even in an idealizing modality—by the contemporary, socially recognized knowledge, it is therefore not worthwhile at all to take that knowledge into account during interpretation, whereas attention should be paid to the context of just that knowledge, sub-sequently—perhaps—socially accepted, which—at least in an idealizing modality—can be attributed to him. We would then be dealing with the 'usual' individualizing concretized inter-pretation. The point is though that very frequently the inter-preter cannot 'subsume' the innovative, subjective context of a performed action under some socially recognized knowl-edge. For the knowledge may not have managed to become established, and moreover, it possibly never will be established. Yet the interpretative application of the assumption of ration-ality unconditionally requires appeal to a socially recognized subjective context. This is not, moreover, a purely doctrinaire requirement resulting from a priori insistence on the concep-tion of humanist interpretation assumed here. It can be illu-strated with numerous examples from research in the huma-nities that intersubjective presentation of an innovative point of view engaged in a given (performed) action is based on con-trasting it with that which constitutes the 'usual', socially func-tioning context of its equivalent in the form of an undertaken action. This is a method frequently used in the humanities. Hence, when a performed action is characterized by a subject-ive context that is not functioning (maybe only for the time being) socially, only correctively concretized humanist inter-pretation comes into the picture.

This situation is most glaringly obvious, perhaps, when we are dealing with an individual interpreter who quite simply is also the subject of the performed action. The correctively con-cretized humanist interpretation involved in that case is a kind of autointerpretation, in the course of which the innovator assumes an attitude toward the existing mental inventory.

A special variant of the situation sketched above is the case in which the set of beliefs composing this innovative knowledge —the subjective context of the performed action—stands in a relation of essentially corrective correspondence to the so-cially recognized knowledge that makes up the subjective context of that action's equivalent (in the form of undertaken action). This is the very case which—as was announced above—we are about to examine more thoroughly.

In the context of this particular case, we are involved with an undertaken action A, to which A' corresponds as a performed action. The idea of the first action is defined by the methodological-theoretic humanistic coefficient associated, shall we say, with knowledge K, while the idea of the second is defined by the coefficient associated with knowledge K'. An analogous situation holds with regard to the effect (sense) E of action A and the effect (sense) E' of action A'. In accordance with the terminological conventions resolved in the preceding chapter, we will say that A' is the correspondence rendering of A, as is E' with respect to E.

Because K' stands in a relation of essentially corrective correspondence to K, we recall that the statement of dependence of E on A holding within the framework of K has no representation in K' in the form of a simple correspondence translation affirming the dependence of E' on A'. It does on the other hand have a representation in the form of an essentially corrected translation. This translation affirms the dependence of E' on a certain subrange of A' (on a certain subset of the situations composing the denotation of idea A' in the framework of K'), and furthermore this subrange correspondingly relates to a certain subrange of A (a subset of the situations composing the denotation of idea A in the framework of K). A case of this sort will be indicated by means of the statement that A' and E', respectively, are essentially corrected correspondence renderings with regards to A and E. What does the relationship between the members of these pairs therefore consist in ?

First of all, a situation of type A', and therefore a situation belonging—in the domain of K'—to the denotations of the corresponding idea (defined in the framework of K'), exists if and only if it is identified in the domain of K as a situation of type A, and therefore as a situation belonging—in the domain of K—to the denotations of a certain other idea (even though that one may well be expressed by the same term as is the idea denoting A' in the domain of K'). Secondly, a situation of type E', and therefore a situation belonging—in the domain of K'—to the denotations of the corresponding idea (defined in the framework of K'), exists if and only if it is identified in the domain of K as a situation of type E, and therefore as a situation belonging—in the domain of K—to the denotations of the respective other idea. Thirdly, it is asserted in the framework of K that whenever A arises, E also takes place, while in the context of K' it is asserted that whenever a certain variant (or subrange) of A' arises, E' also takes place.

Thus, for example, the action of 'dropping' a physical object from a definite height in the context of Aristotelian physics is in any event an action involving the 'setting in motion of the weight' of that object. The correspondence rendering, an essentially corrected one, of that action—which we will designate as A—is, from the viewpoint of Galilean physics, the action A', understood as the surrendering of the given object to the gravitational force of the Earth. From that perspective, action A is only undertaken, whereas action A' is performed. An analogous correspondence holds between the effects (senses) of these actions : E and E'. These entail the given object finding itself on the surface of the Earth after a given lapse of time, respectively, as a result of the operation of the object's 'weight' and as a result of the operation of the gravitational force. Of course, every concrete situation identified in the context of Galilean physics as an action A' (subsumed under the idea A') is identified in the context of Aristotelian physics as an action A (subsumed under the idea A), and *vice versa*. The same is true of effects E and E'. From the standpoint of Galilean physics, it is not true, however, that given an object of properly determined kind at a given height, the object will always fall to the ground after the occurrence of A' in the same amount of time, i.e., that effect E' will always take place. It will be true only when A' occurs under conditions of 'normal' air resistance. Only in reference to that variant of A' is it possible to con-

struct an (essentially corrected) correspondence translation of
the Aristotelian statement of dependence of E upon A. Under
other conditions, beginning with the situation of total lack of
air resistance, the falling time differs—to put it most simply—
very significantly from that which could be determined on the
basis of Aristotelian physics.

Let us linger for a moment in considering this example, because in the further course of our study, it will I believe make
a very instructive illustration of the concept of the relationship
of theory to experience that is to be presented in the next part
of this essay.

We are concerned on the one hand with statement S : 'Every
physical object of kind Q, dropped from height H, falls—under
the influence of its own weight—to the surface of the Earth
in time T'. It should be understood that the letters 'Q', 'H', and
'T' appearing in this statement are abbreviations of defined
constant expressions. The statement is an element of a whole
class C of statements that differ from each other only with
regard to constants of the type Q, H and T. These statements
are thus generalizations, each of which tells after what time
lapse objects of a given kind fall to the surface of the Earth
when they are dropped from a given height. We are also concerned on the other hand with statement S', which is the
essentially corrected correspondence translation of statement
S and which proclaims the following : 'Every physical object
of kind Q, dropped from height H, under conditions of «normal»
air resistance R, falls—under the influence of the force of gravity—in (optionally : about) time T'. Statement S' also belongs
to a whole class C' of corresponding statements. However,
whereas the statements of class C comprise 'discrete' generalizations of common practical wisdom, the statements of class
C' are all consequences of the following assertion N : 'Any
physical object x, dropped under conditions of lack of air resistance from height h, falls under the influence of gravitational force g in time t, in accordance with the following functional dependence : $h(x) = g \cdot t(x)^2/2$; t, moreover, calculated

according to the above pattern, grows additionally in proportion to the growth in air resistance $r(x)$ (which is in turn dependent on the kind q of object x)'. Of course, the functions symbolized by '$h(x)$' and '$t(x)$' (as well as '$r(x)$') take on different values for different locations of object x.

We take note of three fundamental properties of the relation of statement S' to S. These properties apply—generally—to all pairs of statements, the first of which is an essentially corrected correspondence translation of the second, and at the same time, to all statement pairs such that the first member of the pair asserts a dependence of effect E' on a certain variant of performed action A', while the second asserts a dependence of effect E on undertaken action A, where A' is an essentially corrected correspondence rendering with respect to A (as is E' with respect to E). The properties are as follows.

1. No 'internal', logical relation holds between S' and S, since the denotations of the ideas A and A' and of the ideas E and E', are fixed by radically different methodological-theoretic coefficients. The action of dropping a physical object of a given kind from a given height conceived of as the action of the 'setting in motion of the weight' of that object is—as we have already emphasized—something wholly different from the action of dropping an object understood as the submission of that object to the gravitational force of the Earth under the conditions of a definite occurrence, including in particular the condition of zero air resistance[17]. Hence, from the standpoint of S', statement S only expresses a certain belief, but does not offer a competitive conceptualization of the given fragment of reality that is logically commensurable with the literal reference of S'.

Particularly inaccurate, therefore, is the widespread opinion whereby statement N is to be an 'inductive generalization' using individual statements from class C (and thus statements of type S) as its premises[18]. If this were the case, then a logical relation would hold (as is frequently presumed, at least implicitly) between S' and S of the type : consequence of inductive generalization (S')—inductive generalization (N)—premise for generalization (S). Let it be added that this property characterizes every pair : given statement—its essentially correcting correspondence translation, regardless of whether or not the dependence of an effect upon an action is asserted in their contexts.

2. S' characterizes in the terms of knowledge K' (to which it belongs) the range of relative practical effectiveness of statement S, that is, the range of relative practical effectiveness of the action undertaken as A (and performed as A'). For it characterizes in terms of K' the conditions under which the undertaking of action A yields the effect E that is predicted on the grounds of knowledge K (to which S belongs) and understood in the framework of K' as effect E'. Namely, the action of dropping a physical object undertaken as A yields the predicted effect E (close to effect E') under conditions of 'normal' air resistance R. This is 'normal' resistance in the sense that on the average, this is the resistance which objects of kind Q (defined, among other things, with regards to shape and density) most frequently encounter. In cases where resistance diminishes to zero or exceeds the value of R (one should conceive of resistance R as being represented by a definite subinterval of values of $r(x)$), the effect E predicted on the grounds of K fails to appear (that is, the falling time is either significantly shorter or longer). To formulate the matter alternatively, it can also be said that S', by means of its literal reference, characterizes or represents the praxis-objective reference of statement S relativized to K'.

3. The fact that statement S was (and continues, moreover, to be) socially accepted as an element of common practical wisdom can be explained, with the help of statement S' (and this is a consequence of property 2). This explanation assumes —in addition to statement S'—the assertion that beliefs of a predictive type achieve social acceptance if their application in the context of praxis of a historically defined domain renders that praxis sufficiently effective. Undoubtedly, praxis in the days of Galileo and earlier that involved operating physical

objects was carried out, in particular, usually under conditions of 'normal' air resistance. (Appropriate experiments involving the removal of air from the interiors of chambers became possible only later on.) From the perspective of praxis of this sort, statements of the type S (as well as actions undertaken in the social-subject context of the knowledge encompassing these statements) were (are) sufficiently effective[19].

As we see, these last two properties (2 and 3) of the relationship of S' to S—we are now treating these symbols as appropriate variables (and thus in a general fashion)—ensure that although no logical relation holds between the two statements, there is nevertheless an extremely substantial bond of an objective-practical nature between them. This fact enjoins us stridently to oppose the opinion of the advocates of 'methodological anarchy', according to whom qualitative changes in the state of scientific knowledge are to proceed in a fashion which can only be described by the formula : *spiritus flat ubi vult*.

Notice also that explanation, which has recently been discussed, is intimately bound to the correctively concretized humanist interpretation of an action undertaken as A and defined therefore by knowledge K (which contains S), but performed as action A', defined by knowledge K' (containing S'). Namely, the object of explanation here is the fact of the social acceptance of a particularly essential component of knowledge K : the predictive belief expressed by S. The subject of the action performed as A' is in a position to validate his 'vision' of the action if : (1) he reconstructs its hitherto socially functioning conceptualization—from the viewpoint of knowledge K (S), and (2) he explains why society has accepted this conceptualization up until now. As we know (cf. preceding chapter), point (2) is accomplished in the praxis of scientific research as an exposition in terms of K' (S') of the range of relatively effective practical application of this former conceptualization (statement S). It goes unnoticed, however (this was also true in the case of Galileo, as I endeavoured to show in the final section of the preceding chapter) that no logical, 'internal' relation between

K' (S') and K (S) is in any sense established by this sort of measure.

3.3. TWO KINDS OF OPPOSITION OF THEORY AND EXPERIENCE: THE RELATIVE AND THE ABSOLUTE

Without further ado, I propose that we now conceive of the traditionally recognized opposition of theory vs. experience in two variants : (1) the generalized and (2) the special. The generalized opposition of theory and experience will be treated as an exact 'parallel' (in a certain sense) to the relation of essentially corrective correspondence. Thus, the generalized opposition has a relativistic character : if we are given two systems of knowledge K and K', then—from the standpoint of this opposition—K represents experience in the light of K', while K' represents theory with respect to K, if and only if at least some fragment of K' stands in a relation of essentially corrective correspondence to K (to the respective fragment of K).

It can be said of the concept of the generalized opposition of theory and experience introduced in this fashion that its usefulness as a tool for historical epistemology can be doubly substantiated.

First of all, with the help of this concept, it is possible to express—in an essentially corrective manner—a general proposition, usually accepted implicitly in epistemological reflection, which states a regularity involving the constant extension of—or in any event, the constant appearing of new elements of—the 'empirical basis' of science. Within the framework of such reflection, moreover, it is sometimes stated even explicitly (e.g., P. K. Feyerabend or M. Hesse) that the 'observational language' of science undergoes perpetual evolution : statements originally of a theoretical character transform themselves later into observational statements. What is more, it is also said (M. Hesse does so—as we indicated earlier in passing) that the very essence of the development of science is in fact this evolution of the 'language of observation'. At the same time, various representatives of individual scientific fields, and of the mathematized nautural science disciplines in particular, as a rule treat previous 'correspondent' (in an essentially corrective manner) research results as 'experimental data', which for this very reason cannot be passed over in future research and which must be accounted for in a corresponsive mode. Spokesmen for the Copenhagen school interpretation of quantum mechanics even directly require the correspondence (characterized in their trademark 'internal' way) of quantum theory with classical mechanics, conceived of by them as the language of description of 'experimental data', even though classical mechanics has an intrinsically theoretic character in relation to the generalizations of common practical wisdom of Aristotelian physics.

Incidentally, it is to these scientific constructs which directly or indirectly correspond in an essentially corrective manner to common practical wisdom and which (in an analogous way) are 'correspondent', in turn, with respect to subsequent developmental stages of scientific research that the concept, mentioned in the first chapter of this book, of specifically scientific practical knowledge should be applied.

Secondly, the relative concept of the generalized opposition of theory and experience has its own special variant which can be defined as an absolute opposition—by analogy to Galileo's treatment of absolute motion as a special variant of relative motion (though it is to be born in mind that this variant is not, as Galileo imagined, identical with the common, pre-theoretic idea of motion). The absolute concept of the opposition of theory and experience comprises a special case of the generalized concept in the sense that it assumes a special constant relativization : to common practical wisdom embodied (possibly in a precisely rendered form) by scientific constructs of the pretheoretic stage. Every system of knowledge exclusively containing (perhaps precisely formulated) generalizations of common practical wisdom will, in this framework, be called (absolute) experience for short, whereas every system of knowledge corresponding directly or indirectly with this experience in an essentially corrective way will be called an (absolutely) theoretic system or, simply, theory. The absolute concepts of theory and experience thus defined constitute, in turn, essentially corrective renderings of the members of Carnap's opposition, but to a certain degree, they also take into consideration the reservations concerning that opposition that were raised by Quine.

Let us develop this statement somewhat more fully.

Notice first of all that in the framework of the association of interest to us here : (1) Carnap's knowledge formulated in the extended observational language matches a definite region of common practical wisdom, (2) Carnap's theory matches a system of knowledge corresponding in an essentially corrective way to the given domain of common practical wisdom (for simplicity's sake, we will presume that only direct correspondence comes into play) and (3) Carnap's system of correspondence rules matches a set of statements demonstrating that a relation of essentially corrective correspondence holds.

Before we examine all three of these matchings in turn, we have to take note of a general observation. The assumption behind Carnap's characterization of the three fundamental factors just enumerated (knowledge formulated in the extended observational language, theory, system of correspondence rules) as well as the respective relations among them is one of methodological individualism and total ahistoricism. Methodological individualism, with its *a priori* requirement for the reducibility of social phenomena into individual phenomena, in particular dictates to the author of *Testability and Meaning* an understanding of the extended observational language and especially of—its basic component—observational predicates whereby that which he calls knowledge belonging within the framework of that language is in fact a set of particular beliefs assimilated by the individual from the domain of common practical wisdom[20] through the active participation of his psycho-physiological perceptual apparatus. Placing emphasis on this latter factor automatically gives rise to ahistoricism as far as experience is concerned, since the psycho-physiological apparatus does not after all undergo an evolution parallel to changes of the socio-historical type. This ahistoricism is also linked to a synchronic-logical treatment of relations of an essentially (from the viewpoint of historical epistemology) extra-logical and, what is more, diachronic-developmental nature. Of primary concern here is the relation of essentially corrective correspondence expressed by Carnap through the medium of a system of correspondence rules, which logically (syntactically, in this case) connect theory with experience.

Thus, it can generally be stated that the praxis-objective reference of Carnap's conception of the opposition of theory and experience should be placed—from the viewpoint of historical epistemology—in the domain of phenomena composed of the process : (1) assimilation by the individual of common practical wisdom, (2) individual utilization of this wisdom, (3) assimilation by an individual researcher of a scientific-theoretic conception, and (4) the relating (by this researcher) of that conception to the social practical knowledge already assimilated. This process is treated in a socially ahistoric fashion. After all, it is silently assumed that the process runs the same course under all socio-historical conditions. Essentially then, the praxis-objective reference (in the light of historical epistemology) of Carnap's concept of the opposition of theory and experience constitutes only a certain, individualized aspect of that opposition as it is 'seen' by historical epistemology, which considers science (whether in the form of a type of social praxis or in the form of social consciousness) as an anti-individualistically understood social phenomenon that is historically variable in a manner subject to definite regularities.

Let us proceed now to the particular matching factors noted above, beginning with the association : knowledge formulated in the extended observational language—common practical wisdom.

As has been pointed out, the basic distinguishing feature of knowledge belonging within the framework of the extended observational language is the fact that all the descriptive predicates by means of which this knowledge is verbalized are either (1) primitively observational or (2) 'definable' wholly or partially—through the use of these primitives. In a correspondence translation, this would mean that every individual assimilating common practical wisdom or utilization thereof, appealing to his psychophysiological apparatus, learns how—or already knows how—to recognize individual elements of socially functioning denotations of predicates applied to definite actions or their effects. To speak more precisely, we also ought here to include definite circumstances of actions being undertaken. In order to simplify the presentation, we have not been mentioning these. (As we recall, it was established in the second essay that the word 'action' will be treated as an abbreviation for the phrase 'action together with its accompanying circumstances'.) Of course, from the standpoint of historical epistemology, the observational status of Carnap's predicates cannot constitute a criterion designating, by definition, a region of common practical wisdom. It can only be said that this observational status comprises a necessary condition for the utilization by particular individuals of this wisdom and that taking observationality into consideration comprises a point of departure for psychological research also of interest to this epistemology, such as, let us say, research into the question of how it comes about that particular individuals agree among themselves upon the predication of particular predicates of the kind relevant to us here. However, the question of what it is that these individuals experience when they predicate a given predicate in concert and whether their sensations are similar is not crucial for historical epistemology. Whatever they experience in making use—perhaps in an individual, unique manner—of their 'immediate data', the fact is that in using the given predicates in a manner—roughly—in accord with the denotations vested in them by society, denotations that encompass certain actions as well as circumstances and effects of actions, they do not in this way designate these denotations, but—quite to the contrary—they 'accommodate' their own manner of speaking to the form of the denotations already established by society.

In speaking of common practical wisdom so far, we have been working with a definition in the light of which this wisdom comprises a set of beliefs from the sphere of social consciousness which are : (1) general conditional sentences linking definite actions under definite circumstances with their effects, (2) deductive premises for beliefs of type (1), and (3) logical consequences of beliefs of types (1) and (2). Moreover, these actions, circumstances and effects are directly discernible (tangible) in practice. Before undertaking further analysis of Carnap's opposition of theory *vs.* experience, the time has now come to render the above definition more precise in a certain respect.

What is the sense of the qualification 'directly' discernible (tangible) in practice—after all, this is of primary concern here —when used in reference to appropriate situations (actions, accompanying circumstances and effects of actions) ? In any event, one should understand this qualification as placing a condition on the type of knowledge that makes up the methodological-theoretic humanistic coefficient of the literal denotations of the predicates referring to those practical situations. For this knowledge, while defining the given denotations, simultaneously defines and expresses the means of recognizing individual elements of the denotations. It is this knowledge, of course, that forms a certain region of common practical wisdom. After all, it is on this basis—in the case concerning us here— that we name the actions, their accompanying circumstances and their effects which we recognize by means of the appropriate predicates. However, it is clear that we cannot have recourse to the notion of common practical wisdom while characterizing the humanistic coefficient under consideration, because this is the very notion that we are now seeking to define. For this reason we appeal here to a circumstance which we know anyhow must be an attribute of all pre-theoretic knowledge (and therefore of common practical wisdom). This is its—shall we say—'pre-critical' status, i.e., the fact that it does not correspond in an essentially corrective way with any other system of beliefs. Incidentally, such an understanding of the term 'pre-critical status' relates to the meaning bestowed on the word 'critique' by Karl Marx when he qualified his theory of the capitalist means of production put forward in *Capital* as a 'critique of political economy'. It does not seem subject to doubt that this is a critique of previous political economics—involving its essential corresponsive correction.

We are going to identify the condition of the direct discernibility in practice of certain actions, their accompanying circumstances and their effects with the condition of the pre-critical status of the knowledge constituting the social-subjective humanistic coefficient of the denotations of the predicates referring to the respective actions, circumstances and effects. Direct discernibility in practice, understood in this way, is the equivalent herein proposed for the traditional understanding— Carnap's, in particular—of observational status (of predicates) and observability (of traits or relations denoted by the given predicates).

Notice that the above criterion of direct discernibility in practice of elements of denotations of given predicates, which can also be termed the criterion of the direct applicability of these predicates to practical situations places within the sphere of common practical wisdom individual predicates which verbalize that wisdom—in a manner quite different from the traditional criterion for observational status. Thus, for instance, until Carnap's concept of the extended observational language arose, it was commonly accepted within the sphere of influence of logical positivist ideas that the so-called dispositional predicates, such as 'x is soluble in water' or 'x is an acid', do not have an observational character, but rather a theoretical one. The guiding intuition here was the thought that it is not directly, immediately 'visible' whether a given substance is soluble in water or whether it is an acid. Rather, it is necessary to arrange certain conditions or to await their spontaneous emergence (the submersion of a given substance in water, the dipping of litmus paper into a given substance) in order subsequently to be able (indirectly) to infer the appropriate dispositional property. However, on the grounds of the extended observational language, all dispositional predicates were acknowledged 'wholesale', so to speak, as observational. Yet from the standpoint of the criterion of direct applicability of predicates to practical situations, this question is more complex. It is not relevant here whether something—and what specifically—is 'directly visible'. This circumstance may be (and certainly must be) important for the acquisition by particular individuals of competence in using these predicates as well as for the constant maintenance of this competence. On the other hand, what is relevant from the point of view of our criterion is whether or not the social denotations of the pertinent predicates are qualified by pre-critical humanistic coefficients. From this standpoint, the predicate : 'x is soluble in water' is directly applicable in practical situations (as this is colloquially understood). For the point is not whether or not in socially defined (by means of the humanistic coefficient of its denotation) ascertainment of a concrete instance of solubility, we must first of all complete a certain course of action and only subsequently draw the proper inference, but rather that in this ascertainment, we do not make use of any knowledge which corresponds in an essentially corrective fashion with another definite set of beliefs. Quite the opposite situation arises, however, with the predicate : 'x is an acid', understood in the spirit of modern chemistry. The socially functioning literal reference of this predicate is defined, after all, by chemical knowledge that corresponds in an essentially corrective way to matching knowledge from the field of common practical wisdom. To this day even, in some social spheres and situations, knowledge is recognized, on the grounds of which the qualification : 'x is an acid' means the same thing as : 'x has a sour taste'.

As can be seen from these examples, the criterion of direct applicability to practical situations is not as simple in use as the criterion of observational status. Yet, on the other hand, it avoids many hopelessly unresolvable problems that arise in connection with various efforts to specify precisely the—in the final analysis—individual-psychological notion of observational status and observability. As a matter of fact, no one has yet produced a satisfactory answer to the question of what a directly observable property or relation actually is. Of course, a main reason why working with the criterion of the direct applicability of relevant predicates in practical situations repeatedly requires quite elaborate investigations is the fact that on the whole, suitable historical information concerning the development of a given science is indispensable. Moreover, it could not be otherwise—since this criterion is formulated on the basis of historical epistemology. Nevertheless, it can be stated in advance that there do not exist general relations for which the appropriate predicates, traditionally defined as dispositional, or for that matter as introspective, could not be directly applicable to practical situations. Similarly, moreover, it would be difficult to preclude the possibility that among predicates traditionally defined as primitively observational, there might be some predicates which are not (in spite of everything) directly applicable to practical situations.

Let us return now to Carnap's general characterization of knowledge belonging within the framework of the extended observational language. We have established generally that from the standpoint of historical epistemology—correcting the conception in question in an essential way—this is in fact a characterization of knowledge assimilated by the individual from the sphere of common practical wisdom. Thus, to what extent might we be able, from this standpoint, to explain the fact that this knowledge is to be characterized—according to the quite widely held views expressed by the author of *The Methodological Character of Theoretical Concepts*—by the following properties: (1) the 'naturalness' of the empirical semantic interpretation of descriptive predicates verbalizing that knowledge, (2) its realist character, and (3) irrevocability?

Re (1). The fact that the denotations of 'observational' (directly applicable in practice) predicates are defined in a pre--critical way, i.e., by means of a humanistic coefficient that is made up of knowledge belonging to a region of common practical wisdom, evokes an impression of 'naturalness'. The individual operating with this coefficient simply does not notice it, because in no way does it contrast with his remaining common-sense, socially acquired beliefs.

Re (2). The 'transparency', discussed above, of the common--sense humanistic coefficient of literal references generates an illusion of face-to-face communion with objective reality. The illusion is rendered more lasting in that, in the field of activities of directly-practical, 'survival' purposes, common-sense knowledge is as a rule reliable.

Re (3). It is just this assumption of the perspective of common practical wisdom that causes us to treat it as irrevocable. On the other hand, with regards to those features which Carnap, in keeping with the respective widely held views, attributes to (a) theory and (b) the corresponding relation linking theory to common practical wisdom, I submit the thesis that these are opinions expressing the viewpoint of an individual thinking pre-theoretically and confronted with scientific output that is theoretical in the rigorous sense of the word, and hence —the viewpoint of research praxis of the pre-theoretic stage, which nevertheless takes the fact of a scientific theory's existence into account. This thesis explains—in my view—Carnap's characterization both of factor (a) and of (b), and in particular, it explains how it is that this characterization is superficial and 'naive' enough so that it cannot constitute a predictive basis of an effective program for theoretical break--through's in science.

Thus (as has already been suggested), Carnap's conception of scientific theory can—from the perspective of historical epistemology—be located with regard to its praxis-objective reference in the sphere of the individual process of assimilation of a scientific-theoretical concept by a researcher who represents the pre-theoretic manner of thinking and science's pre--theoretical methodological consciousness. Because of this, he is only in a position to discern as much within this theoretical concept as his manner of thinking permits. Indeed, it permits the following features of scientific theory to be preceived : (1) its 'conventionality' ('unnaturalness'), (2) its instrumental (non-realist) character, and (3) its revocability. Of course, if the image of the world provided by common practical wisdom is spontaneously treated as 'natural' and is identified with reality itself, features (1) and (2) of scientific theory must consequently be accepted. If, in keeping this perspective further, the fact, difficult to overlook, of the appearance of successive 'scientific revolutions' in the theoretical stage of scientific development is noted, then there remains only a general impression of permanent revocability (feature (3)), which can be justified only instrumentally. It is clear that basing oneself on this kind of knowledge of scientific theory, it is not possible to formulate any effective program for theoreical break-throughs, which must in any event subjectively assume a point of view that makes appeal to a realist conceptualization of the theoretical image of the world—as opposed to the 'appearance' (in Karl Marx's sense) finding expression in common practical wisdom.

One more supplemental observation is necessary here. The 'spirit'—figuratively speaking—of a researcher representing the pre-theoretical manner of thinking who at the same time perceives phenomena that represent the theoretical stage, a 'spirit' expressed in particular by R. Carnap, is not only conditioned genetically by the societal methodological consciousness of science, which remains within the framework of that first stage, but also functionally—by the need to maintain the world view associated with the pre-theoretic 'vision' of reality. The fact that, in its own way, this 'spirit' takes features characteristic of science of the theoretical stage into consideration may render the researcher able to verbalize functional beliefs regarding certain parts and/or aspects of research praxis from that stage as well—and therefore beliefs that are effective in subjectively motivating the activity that forms those parts and/or aspects.

In the case of the conception of the author of *The Methodological Chaaracter of Theoretical Concepts*, it can be stated that the phenomenon of the emergence of a special kind of theory in the field of the mathematized natural sciences is noted in a most adequate fashion. These theories, after all, cause the least trouble for the positivist mode of thinking, enriched with the idea that not every statement of empirical sciences is translatable into appropriate sentences within the domain of the extended observational language. The term 'theory' is used here in accordance with the terminological custom both of representatives of mathematized natural sciences and philosophers of the instrumentalist disposition. From the point of view being presented here, it would be more adequate to employ the notion of quasi-theories.[21] We will use the latter term now to make a few observations which I believe will permit us somewhat more closely to determine the sense of these last few assertions[22].

As an example which seems to be quite instructive, let us consider the following situation : suppose we are dealing with Newton's law of general gravitation or, alternatively, with Coulomb's law. Both these laws fall under the same formal schema. This can be obtained by replacing functional constants, representing definite quantities, with appropriate variables in the law of general gravitation (or in Coulomb's law). The schema includes all such functional dependence involving a mathematical measure (scalar or vector) of a certain quantity calculated by taking the product of the mathematical measures of two (other) quantities divided, in turn, by the square of the mathematical measure of yet another quantity. We will treat a schema of this kind, defined substantively at most with respect to the relevant quantities' mathematical type, under which certain determined functional inter-dependences of previously defined quantities fall or are believed to be able to fall, as a special instance of a quasi-theoretical statement. A generalized characterization of a quasi-theoretical statement, however, contains a formal schema, under which fall established functional inter-dependences of defined quantities, as well as a schema from which—as an effect of performing appropriate mathematical transformations upon it—a formula with a property of interest to us can be obtained. That is, upon replacing the variables in the formula with the proper constants, representing definite physical quantities, we obtain a statement establishing the respective substantive functional dependence. It is my view that statements of this kind (quasi-theoretical), as well as whole systems of them (quasi-theories), were treated by F. P. Ramsey and afterwards by R. B. Braithwaite as the paradigm for theoretical statements and, respectively as the paradigm for scientific theories (formed by empirical disciplines), since they both assume (as do numerous other representatives of Anglo-Saxon philosophy of science) that a scientific theory comprises statements containing, in addition to logico-mathematical constants, only predicative or functional variables—symbolizing any descriptive constants of appropriate type.[23] This kind of concept of scientific theory is fully compatible with Carnap's instrumentalism : theoretical statements do not assert anything about reality since they operate exclusively on variables replacing descriptive constants. On the other hand, they constitute a logico-mathematical tool for ordering ('experimental') statements establishing definite functional dependences. These are dependences which Carnap would say are stated in the extended observational language.

The fact of the very frequent use made in research praxis, especially that of the mathematized natural sciences, of quasi--theoretical statements and quasi-theories by no means refutes the conception I am proposing here—of a relative vs. absolute opposition of theory and experience. At most it can be said that this fact, in a certain way, complicates the picture regarding this oppositon. The picture presents itself much more simply when we confine our consideration to such theories as Galilean mechanics, Newtonian mechanics, the special or the general theory of relativity, or—to reach into a different field—the Marxian theory of the capitalist mode of production. The question of what special form our opposition assumes in the instance in which we deal with quasi-theoretical statements or quasi-theories is, it is true, profound primarily for epistemological study into the mathematized natural sciences. Nevertheless, in order at least to signal a kind of premise substantiating the opinion that this opposition applied in that arena as well, let us now consider the following.

First of all, a given quasi-theoretical statement can be a formal, logico-mathematical schema of pre-theoretical statements just as well as of theoretical ones. It may even be that the same schema 'services', so to speak, statements of both kinds simultaneously. It can happen, in addition, that a given formal schema is associated with theoretical statements of various 'levels', representing various developmental links in the theoretical stage. Everything depends on the nature of the given substantive statements that come under the respective schema.

Secondly, quasi-theoretical statements and quasi-theories constitute their own kind of provisional substitute for substantive theories, although as 'provisional' substitutes they are sometimes indispensable in practice. In some cases they may, in a heuristically fruitful way, even suggest a direction for investigations leading to substantive theories. For the fact that given substantively defined functional dependences fall under the same formal schema may be a point of departure for the establishment of theory asserting at least one fundamental, substantively defined functional dependence such that statements determining the preceding dependences can even be shown to 'correspond' with it in an essentially corrective way. Of course, however, quasi-theoretical statements or quasi--theories, especially given an instrumentalist understanding of the status of theories, can just as well have a dampening impact on cognitive progress. In the instrumentalist approach, this kind of provisional substitution for a substantive theory is treated, after all, as the end point for research inquiries.

Hence, thirdly, if this instrumentalist tendency to assign scientific theories the status of quasi-theories, a tendency that dampens cognitive progress, is rejected, it ought to be remembered that this progress is defined by the corresponding relation of statements coming under given formal schemata to previous scientific or common-wisdom determinations. What is more pertinent, moreover, is that the relation of essentially corrective correspondence never applies to quasi-theoretical statements or quasi-theories, but exclusively to their defined substitutes. In identifying quasi-theoretic variables with their respective constants (in the context of a given application of a quasi-theoretic statement), we can take these constants from a set of terms verbalizing common practical wisdom—on the foundation of the appropriate social-subjective humanistic coefficient—or from a set of terms already possessing theoretical, socially functioning literal references. It may also come to be that a given substantive substitute for a quasi-theoretic statement comprises only a proposal for a theoretic statement; yet such a proposal still requires the appropriate—primarily corresponding—substantiation.

Thus, Carnap's conception of the opposition of theory and experience is only suited to acting as a basis for a relatively effective program of constructing quasi-theories or quasi-theoretical statements. This, moreover, forms the limit to the range of functionality of that conception in relation to research praxis —including praxis representing the theoretical stage of scientific development. As has been already mentioned, though, it cannot function as a program for theoretical break-throughs. The realist treatment of knowledge from the sphere of common practical wisdom (formulated in the observational language), i.e., the identification of the image of the world presented by that knowledge with familiarized reality, accompanied by an instrumentalist outlook on theoretical knowledge, prohibits Carnap's conception from being able subjectively to guide research activity toward overcoming—in the modality of essentially corrective correspondence—the 'appearance' of common practical wisdom in favour of theoretical cognition. After all, only this kind of activity, correspondingly setting a realistically conceived, theoretical picture of the world apart from common, socio-subjective pre-conceptions about it, leads to theoretical break-throughs in science. In particular, the construction of quasi-theories or quasi-theoretical statements cannot bring about such a break-through. In the situation where the entirety of existing knowledge is represented by the pre-theoretic level, quasi-theories and quasi-theoretical statements introduce only a logico-mathematic deductive organization to the set of pre-theoretic determinations. This organization, however, in no sense changes the status of these determinations. What is more, instrumentalist identification of theory with quasi-theory, it is to be stressed, can have a braking effect on progress in science, because assuming that the attainment of a quasi--theory—and therefore the attainment of a logico-mathematical ordering of a given set of statements—is the same as attainment of a theory suggests the discarding of further quests for real gnosiological progress.

Thus, we have established that the characterization made by R. Carnap of knowledge belonging within the framework of the extended observational language as well as theoretical knowledge and its relation to the former in fact constitutes (from the perspective of historical epistemology) an approximate and idiosyncratically distoreted image of the process of assimilation by the individual of common practical wisdom and of the individual's utilization of that wisdom, as well as an image of how a scientific theory is understood by an individual representing the pre-theoretical point of view. This is a viewpoint which within certain limitations can effectively provide a subjective motivation to the construction of quasi-theories or quasi-theoretical statements. Yet it is thoroughly useless and even a disturbance when it comes to constructing theories that correspondingly 'break through' the cognitive perspective of common practical wisdom or, more generally, when it comes to constructing theoretic systems representing successive links in the chain of cognitive progress. Such an understanding of Carnap's conception of the opposition of theory and experience, i.e., such a presentation of its praxis-objective reference, therefore constitutes a sketch of the corresponding bond (in an essentially corrective modality) to that conception from the position of historical epistemology and in particular from the viewpoint of the alternative construction of that opposition which I have proposed herein.

As we have seen, W. V. Quine stands quite radically in opposition to the concepts of R. Carnap. We will therefore seek to criticize Quine's position correspondingly and thereby be able subsequently to compare it with Carnap's conception—though no longer on the literal plane, that is, on the plane of 'internal', logical relations holding between these two points of view, but rather from the perspective of their praxis-objective references. We will take this task up in the next chapter.

NOTES

[1] W. V. Quine, 'Two Dogmas of Empiricism', in *From a Logical Point of View; 9 Logico-Philosophical Essays*, Harvard University Press, Cambridge, 1964, pp. 40–41.

[2] As a matter of fact, it would be proper to use quotation marks here, since one might point out that the application in this context of the term 'definition' constitutes, from the formal-logical standpoint, a terminological abuse. Yet because we are presenting the ideas and the position of R. Carnap here, quotation marks could be misleading (i.e., the suggestion would arise that these were his diacritics).

[3] Concerning the phrase 'partial semantic interpretation', cf. the preceding note.

[4] Not necessarily, however, radically behaviourist. Some 'intermediary variable'—characterizing the psycho-physiological state of the observer may intervene between these features and the fact of predication.

[5] Remember that as Carnap intended, this idea has nothing in common with the relation of correspondence between systems of scientific knowl-

edge. As a matter of fact, though, a relevant correlation does arise here (the synchronic-logical relation of theory and experience in accordance with the diachronic-developmental relation of pre-theoretic knowledge or current practical wisdom to theoretical knowledge), but this question will not be taken up here, lest this study should become too complex.

6 R. Carnap, *The Methodological Character of Theoretical Concepts*, Minnesota Studies in the Philosophy of Science, Vol. 1, ed. by H. Feigl, M. Scriven, Minneapolis, 1959, p. 69. 'L_T' of course, symbolizes the theoretical language, while 'L_0' and 'L'_0', respectively, symbolize the strict observational language and the extended observational language.

7 *Ibid.*, p. 70. As we see, for the author of *The Methodological Character...*, colloquial 'metaphysical' notions find continuation in scientific theoretical ones.   .

8 W. V. Quine, 'Ontological Relativity', in *Ontological Relativity and Other Essays*, New York-London, 1969, p. 48.

9 W. V. Quine, *Two Dogmas of Empiricism*, cited above, pp. 42–43.

10 *Ibid.*, p. 43–44.

11 *Ibid.*, p. 44.

12 J. Kmita, 'Interpretacja humanistyczna a wyjaśnianie funkcjonalne' ('Humanist Interpretation and Functional Explanation'), in *Elementy marksistowskiej metodologii humanistyki* (*Elements of Marxist Methodology in the Humanities*), ed. by J. Kmita, Poznań, 1973.

13 I am presenting a report here mainly (though not exclusively) on the position cited in the preceding note.

14 It is possible, however, to provide the remaining types of assumptions of rationality, i.e., involving conditions of risk or conditions of uncertainty, with such a form that they will constitute special cases of a specifically generalized assumption of rationality for conditions of certainty. Namely, both under conditions of risk and under conditions of uncertainty, individual actions are assigned definite sets of possible results (of specific probabilities—in the case of conditions of risk). By identifying the sense of each action with the logical disjunction of results belonging to the set assigned to that action, we can then respectively apply two variants of this generalized assumption of rationality for conditions of certainty.

15 I sometimes encounter the question of why I operate (chiefly) with the term 'sense' of an action rather than being satisfied with the term 'goal'. In fact I do so because the idea of goal generally speaking has an individual-subjective nuance, while the idea of sense is much more strongly associated with a value of subjective-social nature, a value linked, moreover, in a subjective-social fashion with the given action. The goal of an (uttered) expression, for example, is on the whole spon-

taneously understood to be a certain intention consciously assumed by a concrete individual, whereas the sense of an expression is rather conceived to be dependent on socially functioning language rules, regardless of how it may be individually adopted. Hence, I operate with the term 'sense' in order to take advantage of the above indicated semantic intuitions that are linked to it, and thereby to facilitate understanding, especially when the question under consideration is directly connected with the concept of humanist interpretation and when the social-subjective significance of the term also comes into focus.

[16] K. Marx, Capital, a Critique of Political Economy, Vol. 1, New York, 1967, pp. 166, 167.

[17] The fundamental difference in the motion of an object dropped in the Aristotelian understanding as opposed to the Galilean understanding is also worth bearing in mind. We analysed this difference in the preceding chapter.

[18] In a typical manner, this opinion is expressed by R. B. Braithwaite in the book Scientific Explanation, Cambridge, 1953.

[19] In the majority of cases, common practical wisdom and the pre--theoretic physics of Aristotle (which remain fundamentally within the sphere of that wisdom) do not so much contain statements of type S as they do certain, less rigorous contrastive generalizations of the sort: 'Every stone falls faster than do wooden objects.', 'Every wooden object falls faster than does an unwound piece of fabric.', etc. Nevertheless, rigorous renditions of such generalizations in the form of statements of the type S are at least potential components of physics on a pre--theoretic level. For as we recall (cf. first chapter), every science in a pre-theoretic stage orders and rigorously specifies (the example at hand is an illustration of this rigorous specification) just such generalizations of the common practical wisdom. Hence we are fully entitled to use examples of statements of the type S, regardless of whether such statements were explicitly formulated sufficiently often.

It is worth noting that generalizations of the type S, precisely because, as has already been pointed out, they co-occur independently, whereas their essentially corrected correspondence translations of the type S' can be deduced from theorem N upon taking complicated measurements of air resistance into consideration, have a more stable, 'solid' character than the latter. They are much less prone to questioning or modification. This state of affairs, which is more and more vividly pronounced as one proceeds to the increasingly 'theoreticized' theorems of physics, sometimes evokes nostalgic sighs from representatives of that specialization. H. R. Post ('Correspondence, Invariance and Heuristics', Studies in the History and Philosophy of Science, November 1971) claims, for example, that physics as a science is quite visibly beset with failures while any botany enjoys unquestionable predictive power within the sphere of its applications (p. 219).

[20] A. Pałubicka takes note of this in her analysis of positivist epistemology undertaken in the work *Orientacje epistemologiczne a rozwój nauki* (*Epistemological Orientations and the Development of Science*), Poznań, 1977.

[21] In some of my works I use the term: 'quasi-theoretic historicism' to distinguish the variant of methodological historicism which assumes a specific character of historical laws—though they are conceived of pre-theoretically. I would like to emphasize (in order to avoid terminological misunderstandings) that the sense of this term does not stand in any close relation with the notion of quasi-theory which is about to be introduced.

[22] Of course, however, the phenomenon of a quasi-theory deserves a separate in-depth analysis. It constitutes—in my view—one of the fundamental tasks of the contemporary epistemological theory of the mathematized natural sciences. This task is being carried out in a highly inspiring way by W. Mejbaum.

[23] Cf., for example, R. B. Braithwaite, *Scientific Explanation*, Cambridge, 1953.

CHAPTER 4

THE DUHEM-QUINE THESIS

4.1. THE COMPREHENSIVE INSTRUMENTALISM OF W. V. QUINE

As we noted in the preceding chapter, W. V. Quine rejects every construction of the nature of Carnap's concept of an observational language (either strict or extended) as substantively incorrect since—in his opinion—it is based on the false assumption that the vocabulary serving to verbalize our knowledge about the world also includes predicates, elsewhere referred to as primitively observational, each of which is equipped, independently of all other predicates and therefore in a 'natural' way, with a denotation in the form of a definite observable relation. This assumption is false because first of all, there are no such terms in our language which denote objects (in the wider meaning of the word) of one kind or another in a manner totally independent of that which all the remaining defined terms denote. Secondly, the denotations of these terms, in particular of the terms regarded by Carnap as primitive observational predicates, are not assigned to the terms directly and 'naturally', but rather on the basis of a specified set of ontological-semantic assumptions. It is this very set of assumptions that designates the complex distribution, so to speak, of references of individual elements of the lexical system. A distribution of this kind can be carried out in a variety of ways—corresponding to a variety of systems of ontological-semantic assumptions, each of which is capable of accounting for the purely empirical data represented by 'stimulus meanings', that is by types of physically characterized situations determining positive or negative replies to appropriate occasion sentences. Thus, the choice of one of the possible ontological-semantic systems is empirically arbitrary. It is dictated by considerations of a formal-technical nature. This is true both when we acquire our own mother tongue and when we conduct linguistic research into some foreign language.

The fact that the basis embraced by Carnap for distinguishing the observational language from the language of theory—the 'naturalness' of the empirical semantic interpretation of observational predicates versus the conventional nature of correspondence rules binding statements of theory with statements from the domain of the observational language—proved to be illusory renders, as a consequence, the distinction between statements of the observational language and theoretical statements also illusory. The former were supposed (according to Carnap's distinction) to be in principle definitive and irrevocable, whereas the latter were by nature meant to be subject to revocation. However, a conventional aspect always comes into play in establishing references for expressions, and there is no difference in this respect between vocabulary regarded as observational and that regarded as theoretic. Moreover, both these groups of terms, being indistinguishable from each other (until empirically arbitrary ontological-semantic decisions are made), form a uniform lexical system. Therefore, the incompatibility with experimental data which a given system of knowledge may fall prey to can be averted just as well by a change in the semantic interpretation of a number of sentences otherwise regarded as observational, which may entail the repudiation of several of them, as by a change in the semantic interpretation of a number of sentences otherwise regarded as theoretical statements, which of course may also entail the renouncement of a number of these statements. If, in the practice of scientific research, we more frequently encounter the second course of action, it so happens not for substantive-epistemological reasons, but for formal-technical ones. Consideration for the factor of 'economy' is decisive here : economy in the range of inescapable changes and in the degree of complexity both of the course of action itself and of the achieved effect.

Finally, if the application of a criterion for truth to sentences which are not logical truths (R. Carnap would speak here
of synthetic sentences, but as we know, the author of *Ontological Relativity* rejects the dichotomy of analytic and synthetic
sentences) were to make sense only when those sentences are
unambiguously associated in a 'natural' way with given states
of affairs, then both sentences elsewhere reputed to belong to
the (extended) observational language and sentences elsewhere
passing for theoretical statements—in equal measure—fail to
come under this criterion. Sentences of both the first and the
second kind bear an instrumental status. They constitute an
instrument for organizing the purely empirical data—which
are not (by themselves) in a position to determine the truth
or falsehood of any sentential linguistic unit—of our knowledge
about the world. Thus, the last reason also falls through for
distinguishing knowledge formulated within the (extended)
observational language (of experience) from theoretical knowledge : the supposed real status of the former versus the instrumental status of the latter. The entirety of our knowledge
of the world has an instrumental status. '... in point of epistemological footing ... physical objects and ... gods differ only
in degree and not in kind. Both sorts of entities enter our
conception only as cultural posits. The myth of physical objects
is epistemologically superior to most in that it has proved more
efficacious than other myths as a device for working a manageable structure into the flux of experience.'[1]

It is this outlook of W. V. Quine's outlined above—assigning
an instrumental status to the entirety of our knowledge about
the world—that I call comprehensive instrumentalism. (The
term 'total instrumentalism' could also be used here.) Of course,
this outlook ought to be differentiated from the instrumentalism represented by Carnap. The latter embraces only theoretical knowledge, whereas knowledge formulated in the (extended) observational language is concurrently understood in
a realist light. This latter point of view will be called dualistic
instrumentalism—assuming a duality of the cognitive status of

the two respective components of our knowledge about the
world. As we will see later on (in the next chapter), comprehensive instrumentalism is not a concept encountered on an
exceptional basis (even though arguments on its behalf are
quite diverse—Quine's argumentation distinguishes itself by
its originality and inspirational qualities). The adherents of
this epistemological orientation include even some Marxists as
well[2].

As can be easily observed, for the author of *Ontological Relativity*, an immediate premise for comprehensive instrumentalism consists of the idea that no substantive-epistemological
reasons exist for regarding knowledge belonging—from the
viewpoint of advocates of Carnap's conception—to the domain
of the (extended) observational language as not subject to revision guided by extra-empirical motives and thus as less conventional than knowledge which (from that viewpoint) is theoretical. Through assigning modified references by convention
to statements which have hitherto been widely regarded as
observational and consequently renouncing at least some of
these statements, a measure which is sometimes taken in the
research praxis of empirical sciences, we can attain a compatibility of our knowledge with purely empirical data that is
every bit as thorough as that attained by the alternative
measure of revising (supposedly) theoretical statements. For instance, if we simultaneously accept a (supposedly theoretical)
statement concerning the rotational motion of the Earth as
well as the (supposedly observational) statement that a stone
dropped from a tower carries out a motion parallel to that tower, then the incompatibility of these two statements—appearing on the grounds of certain other propositions—can be removed equally well by a semantic re-interpretation of the predication of the rotational motion of the Earth, which would lead to
the renunciation of the first statement, or by such a semantic
re-interpretation of the predication of the motion parallel to
the tower executed by the stone dropped therefrom that the
second statement would consequently be revoked. As is widely

known, Galileo chose the latter option, at the same time maintaining full compatibility with purely empirical data (Quine's viewpoint is still presented here in 'intermediate' reported speech).

The immediate premise outlined above for the comprehensive instrumentalism of W. V. Quine also constitutes the point of departure for the strengthening of the well-known thesis of P. Duhem, a strengthening provided in Anglo-Saxon philosophy with the name : the Duhem–Quine thesis.

As is commonly known, the author of *The Aim and Structure of Physical Theory* formulated the idea (referring principally to physics) that not only is a definitive substantiation for a given theoretical hypothesis impossible, but also—and most importantly—that in the face of the incompatibility of the hypothesis with observational data falsifying it, there is a possible and actually practiced procedure involving the modification or supplementation of statements commonly accepted, in conjunction with which the hypothesis leads to observational consequences contradictory to the data obtained to date. Upon introduction of the relevant supplementation or modification, our hypothesis can still be maintained. For this very reason—in P. Duhem's view—one ought to accept the opinion that there can be no definitive refutation of any (theoretical) hypothesis by any observational determinations (taking note—as he assumed—of observational data in a 'natural' and immediate way)[3].

The above thesis of Dubem is strengthened by the epistemological deliberations of W. V. Quine in the sense that they provide additional reasons why the thesis ought to be accepted. Namely, there is a special kind of statement, in conjunction with which certain empirical data can be deduced from a given hypothesis. It consists of assumptions that help semantically interpret the sentences taking note of those data. A modification of assumptions of this kind may disqualify the (*prima facie*) refuting force of observational determinations. Galileo proceeded in precisely this manner with the observational statements noting the motion of a stone falling parallel with respect to a tower.

The Duhem–Quine thesis evoked numerous protests from philosophers who could not reconcile themselves to its 'nihilist' implications. They did not want to agree that every construct of our knowledge of the world should be empirically arbitrary in so broad a measure that one could not apply to any of those constructs the criterion of truth understood, of course, in the sense set forth above (as a 'natural' compatibility with purely empirical data). Yet the epistemological perspective they embraced did not permit any other understanding of this criterion.

Thus, the revocability of statements elsewhere called observational is a result primarily, according to W. V. Quine, of the fact that like all other statements of our knowledge, they assert definite states of affairs only on the basis of an appropriate set of ontological-semantic assumptions. Thanks to this set, the entire vocabulary articulating that knowledge is equipped with definite references. Since, moreover, the choice of such a system of assumptions from the expansive set of anologous systems, each having the property of allowing our knowledge to agree with the pure data of experience (with the stimulus meanings of occasion sentences), is empirically arbitrary and determined only by considerations of a formal-technical nature, we can avert the emerging incompatibility of heretofore accepted knowledge with new experimental data by a change, in particular, of the system of ontological-semantic assumptions of a kind that would lead to the revision of the references of statements otherwise called observational and, as a consequence, to the renunciation of some of these statements.

It should be stated here that Quine's concept of a system of ontological-semantic assumptions, as well, for that matter, as the concept of 'ideological beliefs' lying at the foundation of P. K. Feyerabend's 'natural interpretation'[1], is probably most reminiscent, among the notions arising within the sphere of Anglo-Saxon philosophy of science concerning this matter, of the characterization that has been proposed in this book of the methodological-theoretic humanistic coefficient of references. From this perspective, the said coefficient can be associated with Quine's system as its correspondence rendering (of an essentially corrective character). The above operation (of association) allows us quite effectively to grasp the objective function of those positions which advocate the Duhem–Quine thesis with various sorts of reservations as well as those which attack the thesis.

It is to be stressed that the advocates of our thesis as a rule do not fully understand Quine's idea. Namely, they do not understand that additional theoretical assumptions of a special kind, attached to a hypothesis being tested, are of prime relevance in this conceptual framework. (P. K. Feyerabend is excepted here of course.) By this I mean the ontological-semantic assumptions, that is, the methodological-theoretic coefficient of references concerning, in particular, the observational statements 'testing' the hypothesis. Even J. Giedymin, an author who so meticulously notes the shades of meaning in particular formulations, observed only that the Duhem–Quine thesis, which he once accepted, asserts 'that we always test a hypothesis by deducing a consequence not from the hypothesis alone, therefore the falsehood of the consequence does not logically force us to reject the tested hypothesis, since we can reject the false assumption.'[5] Furthermore, in claiming to have undertaken a 'logical strengthening' of the thesis and failing to perceive that in reality he was simply approaching its authentic sense more closely, he is content with an assertion of the 'systematic ambiguity of observational results', arising because 'there is no unambiguous logical link between the result of observation, represented by an isolated observational report. and any isolated observational sentence. In particular, an observational report not supported by additional assumptions does not exclude any isolated observational sentence at all'[6]. The point is that Giedymin has replaced Quine's notion of ontological semantic assumptions with a category of assumptions that is much poorer in content and at the same time less universally applicable. These assumptions designate logical connections among theoretical hypotheses, observational sentences and observational reports that either falsify or confirm those sentences. The assumptions can perform such a role due to the fact that they more precisely define (from the theoretical standpoint) the states of affairs asserted by observational sentences. Concerning these states of affairs, it is affirmed in observational reports that they have been noticed. Thus, the assumptions function similarly to Quine's ontological-semantic prejudgments. On the other hand, however, Giedymin does not observe that Quine primarily has just this ontological-semantic function of the prejudgments in mind, and therefore means to say that no sentences at all—not only observational sentences or reports—would have references if it were not for these prejudgments. Hence, he treats the assumptions he has introduced as a kind, in no way extraordinary, of additional premises attached to the hypothesis being tested which make possible the deduction from the given hypothesis of certain reported observational sentences.

Typical opponents of the Duhem–Quine thesis understand its sense even more vaguely. Namely, the standard interpretation of the thesis in this school is expressed in the proposition that for the empirical testing of theoretical hypotheses or of whole systems of theories, it is imperative to rely on certain supplementary assumptions (as additional premises of the testing process). These assumptions cannot be empirically substantiated separately and therefore have a conventional character. In this way, W. V. Quine's ontological-semantic system of reference, which in J. Giedymin's conceptualization was resolved into a group of assumptions precisely specifying the theoretical 'content' of observational sentences (and thereby establishing their logical relation to theoretical hypotheses), here undergoes even more extensive abstraction : it has been transformed into a set of additional assumptions which cannot be separately tested empirically. As can therefore easily be observed, the typical opponents of the Duhem–Quine thesis naturally assign R. Carnap as well to the camp of its proponents, since his system of correspondence rules, conventionally linking a system of theoretical statements with realistically interpreted findings formulated in the (extended) observational language, clearly comprises a special case of those conventional additional assumptions which cannot be substantiated in a separable empirical way. Hence, the diagnosis arises that the opposition, thus characterized, to the thesis under discussion constitutes an expression of the orthodox positivist viewpoint, which rejects the possibility of even the dualistic-instrumentalist form of reconciling positivism with the fact of the emergence of scientific theories.

Of course, not all opponents of the Duhem–Quine thesis come under the above diagnosis. K. R. Popper, for instance, does not agree with the thesis. Concerning the position of P. Duhem, he stated that it is only accurate insofar as—essentially—no theoretical hypothesis can be empirically substantiated in a definitive manner, but that it incorrectly assumes that no theoretical hypothesis can be definitively refuted. In other words, in Popper's opinion, an *experimentum crucis* of positive result is impossible, but a decisive experiment of negative result is possible. Thus, for numerous reasons, an outlook of this kind cannot be regarded as an attempt to defend orthodox positivist epistemology. We will look at two of those reasons. First of all, the supposedly definitive refutation of a theoretical hypothesis can only in appearance be—on the grounds of hypotheticism—definitive, since the observational sentences refuting the given hypothesis are themselves treated here as revocable hypotheses. Secondly, the above 'naive falsificationism', as I. Lakatos qualified the earlier version of Popperian hypotheticism, blatantly incoherent in its conception as a whole, does not at all emerge as a means of defending the realist interpretation of knowledge formulated in observational terms. Quite to the contrary, it is an effect of the tendency to uphold the concept of the realist status of scientific theory. The non-definitive falsifiability of theoretical hypotheses, caused by the conventional nature of the assumptions used additionally during their verification seemed to Popper to be irreconcilable with his thesis of the gradual, steady growth of the 'truth content' of theories appearing one after another in the process of scientific development.

The leader, so to speak, of the orthodox-positivist opposition to the Duhem–Quine thesis, one the other hand, is A. Grün-baum. He presents the thesis, in a way that confirms this diagnosis, as follows. The thesis is to assert that in a situation involving the application of the refutation schema

$$(H \wedge A \rightarrow O) \wedge \sim O$$

'no matter what the specific content O' of the *prima facie* adverse empirical evidence $\sim O$, we can always justifiably affirm the truth of H as a part of the theoretical explanans of O' by doing two things : (1) blame the falsity of O on the falsity of A rather than on the falsity of H, and (2) so modify A that the conjunction of H and the revised version A' of A does entail (explain) the actual findings O'.'[7]

Having interpreted the Duhem–Quine thesis in this way, Grünbaum goes on to distinguish two senses of the thesis : (1) its trivial sense, whereby it is correct in an obvious way, and (2) its nontrivial sense, on the basis of which it is both unsubstantiated and subject to refutation.

Taken in its trivial sense, the Duhem–Quine thesis allows for such possibilities as the performance of certain formal-logical tricks (such as the replacement of A by '$H \rightarrow O'$' or simply by O'). It also takes into account the possibility of radical changes in the way certain terms engaged in the testing of H are understood. Of course, transforming A into A' in a trivial way one can 'save' hypothesis H quite assuredly. For example, in the face of all expected empirical data, we can uphold the hypothesis : 'Buttermilk is a deadly poison for humans', if we attach to the hypothesis an assumption which will render the term 'buttermilk' synonymous with 'arsenic'. It is not subject to doubt, however, that the Duhem–Quine thesis is interesting from the methodological and epistemological point of view only when we do not understand it trivially, and thus only when we place a condition on the operation of modification A that will prevent the use of formal-logical tricks and the introduction of radical semantic changes. And once the sense of the Duhem–Quine thesis is restricted in this way, it can be convincingly shown that the assertion of H (in this restricted, non-trivial manner) comprises a formal and a substantive error.

The formal error of relevance here is one of *non sequitur*. We have no substantiating premises for the assertion that for every H and O', there is a non-trivial A' such that O' follows from H and A'. In fact, each separate instance of the incompatibility of H and A with O' requires its own demonstration of the existence of an appropriate A'. The mere fact of the occurrence of this incompatibility in no way whatsoever guarantees the existence of such an A'.

The substantive error is that our thesis is false from factual perspectives. Namely, specific instances can be pointed out which refute it. In particular, physical geometry is a case in point. As we know, A. Einstein stated, in accordance with the idea of Duhem, that the accuracy of a given system of physical geometry cannot be empirically verified in a separate way. Consequently, the choice between Euclidean geometry and any non-Euclidean geometry cannot be independently resolved empirically. For if we have a given system of geometry, supplemented by a physical interpretation of certain of its concepts, and especially of the concept of congruence, in order to submit the system to empirical verification, we must additionally have recourse to certain laws of physics which stipulate what deformations (thermal, electromagnetic, elastic, etc.) may effect metallic measuring rods being used for the observational determination of the congruence relation. Yet these very laws, indispensable for correctional computations, themselves appeal to a definite geometry. The point now is that those correctional laws, originally formulated on the basis of Euclidean geometry, can be modified in such a way that, upon adjusting them to a particular non-Euclidean geometry, we achieve compatibility of the empirical data with the given geometry. Nevertheless, the outlook of Einstein that we have just outlined—in Grünbaum's opinion—is fallacious; contrary to that outlook, a given system of physical geometry can be submitted to independent verification. This very fact is an instantiation of factual refutation of the Duhem–Quine thesis (non-trivially understood).

Thus, as we see, the point of view represented by A. Grünbaum exemplifies the orthodox positivist opposition to the Duhem–Quine thesis in an especially instructive way. (1) Quine's specific strengthening of Duhem's idea is passed over. That is, the strengthening simply goes unobserved. (2) Argumentation concentrates on the goal of showing that the additional premises noted by Duhem, to which the theoretical hypothesis being tested is coupled for the sake of the deduction of definite observational consequences, can separately be empirically verified.

Among other things, it was stated above that A. Grünbaum— as a model representative, as it were, of the orthodox-positivist opposition to the Duhem–Quine thesis—simply fails to notice Quine's strengthening of the idea of P. Duhem. Rather, for precision's sake, it should be added that it is not the case that he takes no note of it at all, but that he does so in a specific, caricatured fashion. What I had in mind by asserting this oversight of Quine's supplementation of Duhem's conception is that within the context of outlooks such as Grünbaum's, the strengthening acquires an interpretation which so distorts its 'intensional' shape that for practical purposes, it can be regarded as unobserved in that context. We will return to this question shortly. Let us now present in a few sentences the argumentation with which Grünbaum intended to show the independent empirical verifiability of a given system of physical geometry. The point is not to exhaust the plan of presentation of this philosopher's position (regarding the matter of interest to us

here) so much as to reinforce the demonstration of the orthodox-positivist features of this position.

Thus, the fact that a hypothesis of a given system of physical
geometry is subject to separate empirical verification is supposed to be obvious upon consideration of the two following
points.

First of all, there exist regions in time-space in which no
deformations of measuring rods occur. Their existence can be
determined in a simple way : through assertion of the fact
of the ubiquitous coincidence in the given region of two rods
of different chemical constitutions. Moreover, if someone
wanted to regard even that assertion as an ambiguous observational generalization, then his position would disqualify itself
as one reduced 'to the absurdity that any total system can be
espoused as true a priori'[8]. For such a position would assume
that 'no observation could ever possess the required univocity to
be incompatible with an observational consequence of a total
theoretical system'[9]. Anyhow, with respect to those regions of
time-space in which the relevant deformations fail to occur,
we do not have to escape—in the course of empirical verification of a hypothesis of a given system of physical geometry—
to the help of extra correctional physical laws.

Secondly, regions where the given deformations do occur
also enter into consideration of course. Hence, if our geometric
hypothesis is to have universal value, it must be empirically
verified with respect to these regions as well. But contrary to
the views of Einstein, we certainly cannot adjust these laws
however we please, in the spirit of a geometry which would
only permit the reconciliation of the given hypothesis, supplemented by the appropriately adjusted shape of those laws, with
observational data. Only that kind of geometry (or hypothesis
about geometry) comes into play which both defines the type
of empirically applied measurements and—on the other hand—
is indicated by that type. By postulating free choice of physical
interpretation of geometric concepts, we thereby postulate the
possibility of radical semantic changes of the terms expressing

these concepts. This possibility has already been ruled out, however, since it only comes into the picture given the trivial understanding of the Duhem-Quine thesis.

Let us now turn to what is specifically Grünbaum's (orthodox-positivist) interpretation of Quine's strengthening of Duhem's idea. This interpretation, in a manner that is certainly unintended, takes note of this strengthening as a special case of the trivial unedrstanding of the Duhem-Quine thesis[10] : a case in which radical semantic change is assumed for terms appearing in the theoretical hypothesis undergoing empirical verification. For essentially, when W. V. Quine claims that it is possible for a given theoretical hypothesis to be maintained, given a change in our system of ontological-semantic assumptions of such a kind that, once this change has been carried out, the observational statements which were previously refuting the hypothesis now, as a result, assert different states of affairs, he is at the same time undoubtedly permitting—at least implicitly—the emergence of what A. Grünbaum calls a radical semantic change in the terms being used in the theoretical hypothesis. That is, after this transformation of our system, the reference of the hypothesis is changed in such a way as to permit its retention. In Grünbaum's eyes, the above situation resembles someone making the term 'buttermilk' synonymous with the term 'arsenic' and thereby propping up a hypothesis which declares that buttermilk is a deadly poison for humans. What does Grünbaum's farcical distortion of Quine's concept involve (disregarding the fact that the sentence : 'Buttermilk is a deadly poison for humans', as it is normally understood, cannot be regarded as a theoretical hypothesis in either of the meanings of the term set forth in the preceding chapter : either in the relative sense or in the absolute sense) ?

First of all, it is a distortion in that an assertion of the synonymy of 'buttermilk' and 'arsenic' or—adhering precisely to the literal content of this concept of Quine's (who, as we know, questions the very possibility of a valid construction of the idea of synonymy)—an assertion of the form of equivalence :

'For every x, x is a portion of buttermilk if and only if it is
a portion of arsenic' is, according to this conception, only a consequence of the respective change in our system of ontological-
-semantic assumptions, but not the direct expression of a decision for that change. The involuntary *naïveté* of Grünbaum's
interpretation of the concept in question lies in the fact that he
imagines that it suffices to decree that the term 'buttermilk'
be assigned the same reference as the term 'arsenic' in order
for Quine's change in the system of ontological-semantic assumptions to be effected. A clearly discernible influence here is
that of the positivist belief that each of these two terms has
an autonomous, 'natural' semantic interpretation, independent
of the interpretation of the other term as well as of the interpretation of all other terms appearing beside it—in the system
of statements forming ontological-semantic assumptions. It is
sufficient, to take this into consideration, as A. Grünbaum
naively overlooks, in order to appreciate how complicated
a possible change, conceived in accordance with Quine's concept, in the system of ontological-semantic assumptions would
have to be in its course and its effects, if it were to justify the
synonymy cited above.

Consequently, a second aspect of the distortion under consideration here also stands out. Quine's conception precludes—
as impermissible—any such change in the system of ontological-semantic assumptions that would be analogous to the change
producing the synonymy (equivalence) offered by A. Grünbaum as an example. This is because it would transform the
hitherto existing references of many predicates (related *via*
existing ontological-semantic assumptions to predicates of the
type : 'x is a portion of buttermilk', 'x is a portion of arsenic'),
from a formal-technical perspective, in an unusually complex
manner. It does so without assuring any formal-technical 'sav-
ings' for our theoretical organization of experimental data. Such
a change would therefore be wholly 'uneconomical', and this
makes it impermissible.

It must of course be conceded that the criterion, postulated by Quine's conceptualization, of the formal-technical 'profitability' of Grünbaum's radical semantic changes of relevant terms is sufficiently indefinite for it not to disqualify in a readily evident way even the most naive interpretations of these concepts. Only this circumstance makes it possible to understand why Grünbaum's argumentation is quite commonly treated as a genuine polemic with respect to Quine's strengthening of P. Duhem's concept.

4.2. THE COMPREHENSIVE INSTRUMENTALISM OF W. V. QUINE FROM THE VIEWPOINT OF HISTORICAL EPISTEMOLOGY

In the preceding section, we determined that R. Carnap's characterization of the opposition of theory and experience carries on the basic, traditional positivist outlooks to such an extent that it is incapable of forming a methodological program (a set of methodological norms and directives) which could function as an effective socio-subjective context of research praxis passing through a process of changes leading from the pre-theoretical stage toward the theoretical stage, not to speak of the possibility of this program functioning with respect to research which has already been established as theoretical. Yet in the case of the comprehensive instrumentalism of W. V. Quine —the contrepiece of his conception being expressed by the Duhem–Quine thesis (more precisely, by Quine's strengthening of P. Duhem's idea)—this matter presents itself somewhat differently. Namely, in contrasts to Carnap's position that statements constructed within the framework of the so-called extended observational language have, in principle, a definite character, and that in conjunction with this (and by reason of this—according to Carnap), they are characterized by a real status, the author of *Ontological Relativity* assumes the position that these statements are just as subject to revocation as are theoretic statements (in Carnap's sense) and that they possess a cognitive status analogous to that of the latter : instrumentalist. Thanks to this, Quine's comprehensive instrumentalism provides a basis for a methodological program which would permit the construction of systems of knowledge that do not encompass previously accepted, relevant generalizations located in the sphere of the extended observational language, that is—if we apply the conceptual apparatus of historical epistemology—generalizations of common practical wisdom. What is more, this program would not postulate a simple bypassing of these generalizations in the event that they proved to be inconsistent (in Quine's understanding ; as a matter of fact, though, logical incomparability is always involved here) with the new system, but rather would postulate their semantic re-interpretation, brought about through a change in the old system of ontological-semantic assumptions, i.e., through a change—to make use again of a correspondence translation—in the previously operational methodological-theoretic humanistic coefficient. The new coefficient, moreover, comprises the context determining the literal reference of the statements 'being tested', that is, of the theoretical statements—in our understanding (since they are at odds with the generalizations of common practical wisdom)—in such a way that in the light of that reference, the statements are true in the formal, semantic sense of the term.

Because the construction of knowledge systems of this very kind is a basic feature of science in a period of theoretical break-through, it can be asserted that Quine's comprehensive instrumentalism implicitly embraces a methodological program which is functional, to a certain extent, with specific reference to the period of break-through. In other words, this implicit program constitutes a subjective expression of that period.

However, is this program sufficiently functional from the perspective described above ? Two main points make it necessary that we qualify the program as one which to only a small extent can be instructive for research procedures proper to a period of theoretical break-through.

Consider first of all the semantic re-interpretation, postulated by this program, of previously accepted generalizations of common practical wisdom, generalizations traditionally regarded as statements from the domain of the extended observational language. This re-interpretation, which maintains the (formal-logic) truth of theoretical hypotheses, is presented as an operation, the arbitrariness of which is limited exclusively by considerations of a formal-technical nature. However, the program provides no instructive data even roughly suggesting that the point here is not simply to re-interpret certain experimental generalizations, but rather to approach them in a manner that takes into consideration their former methodological-theoretic humanistic coefficient—in the mode of essentially corrective corespondence, which gives an account of the reasons for the practical effectiveness which they hitherto enjoyed as well as the range of this effectiveness. In this respect, Quine's program recalls Popper's program for theoretical break-through, which postulates the possibility of rejecting fundamental statements (as well as generalizations of them) in the event of their incompatibility with theoretical hypotheses, but it provides no data which would indicate that something additional has to be done with these 'rejected' fundamental statements and that only under this condition may they be rejected. Certainly, Quine's program is more instructive than that of Popper on one point : it signals (in its own specific way) that the 'rejection' of previous experimental determinations takes place on the grounds of a new methodological-theoretic humanistic coefficient (on the grounds of a new system of ontological-semantic assumptions). Given this new coefficient, we reject some previous observational constructs, while we retain others, although we now endow them with new literal references (and hence—as a matter of fact—with the status of theoretical statements). Incidentally, that which determines the greater instructiveness of Quine's program also constitutes, just as in the case of Feyerabend's concept of the 'ideology of natural interpretations', the source of epistemological scepticism or—respectively—of 'methodological anarchy'. For if the operation of changing a methodological-theoretic humanistic coefficient, coupled with a specific 'invalidation' of previously accepted experimental determinations is to be so fully arbitrary, then this arbitrariness must also extend to the very procedure for constructing and accepting scientific theories. (As we recall, A. Grünbaum speaks here of the 'absurdity' of the *a priori* truth of any sufficiently elaborate theoretical system whatsoever.) Popper's program avoids this consequence on account of its assumption (not conscious, of course) of the invariability of the methodological-theoretic humanistic coefficient.

The second reason for this state of affairs, that Quine's program for theoretical break-through is instructive only in a limited way, boils down to a consequence of the understanding, within the context of this program, of the change of a methodological-theoretic humanistic coefficient as a change that is arbitrary with respect to experimental determinations, and of (the understanding of) the traditionally positivist idea that if a given statement cannot be unambiguously and 'naturally' qualified with regard to its truth value, then it can't possess any truth value at all. The consequence of this, as can be easily seen, is a totally instrumental conception of scientific statements : neither constructs elsewhere referred to as observational nor constructs elsewhere referred to as theoretical can lay any claim to the status of more or less adequate descriptions of objective reality. Yet the fact of the matter is that the subject of a transformation taking place in research praxis—a transformation from the pre-theoretic stage to the theoretic—cannot uniformly assign an epistemological value to the generalizations of common practical wisdom, which are correspondent in an essentially corrective manner, as well as to the theoretical statements corresponding to them. He can't act in this way because such a posture would preclude an effective realization of the corresponding operation. A 'seriously' (realistically) regarded scientific theory must answer the question of what it is in the reality characterized (literally) by the given theory that has brought it about that the correspondent generalizations, considered together with the original (and now no longer 'seriously' regarded) methodological-theoretic coefficient, were able, within a definite domain of this reality, to constitute so-cio-subjective premises for sufficiently effective activity. Thus, a realist conception of 'one's own' theoretical statements is an indispensable condition for the respective research operations to be able to participate in the transformation of scientific research from the pre-theoretical stage to the theoretical. From this perspective, a more instructive program is that of P. K. Feyerabend, who—in a purely normative manner, it is true—requires a realist interpretation of scientific theories, as does K. R. Popper, moreover.

In speaking here of the exclusively normative argumentation of P. K. Feyerabend in favour of a realist conception of scientific theories, I have in mind the fact that this author does indeed motivate this requirement by indicating that its realization is indispensable for cognitive progress in science, but this progress is understood purely axiologically : as a value unto itself. Incidentally, this is strongly reminiscent of a value postulated by representatives of the Frankfurt School, which they termed the emancipation of the individual. This value involves releasing the individual from the 'fetters' of erroneous consciousness, from 'ideology'. This, as we see, exhibits an obvious analogy with Feyerabend's release of the individual from the existing 'ideology of natural interpretations'.

From the perspective of historical epistemology, the term 'cognitive progress in science' possesses, in addition to its gno-siologico-axiological sense, a definite descriptive sense. The fact that this progress truly occurs, through the constant reiteration of the relation of essentially corrective correspondence, can be illustrated without the intervention of any strictly normative premises, as can the fact that a realist conception of newly proposed theories is subjectively imperative for that progress. Of course, it is also evident from this perspective that this realist conception, proper to the research-practitioner, can never be adopted by the epistemological theoretician in its absolute form. To a given historical theoretic truth, he can only ascribe a value of relative truth, occupying a definite place in the sequence constituting cognitive progress. Therefore, if he accepts the perspective under discussion, he will regard the absolute identification of the theoretic image of reality with objective reality (with the praxis-objective reference of the given theory) as an objective-idealistic gnosiological error which the philosopher (the epistemological theoretician) is not allowed to make, for purely epistemological reasons, but which at the same time may, for the research-practitioner, comprise a viewpoint that promotes cognitive progress at the specific historical moment, even though by petrifying himself and petrifying his attitude of acceptance toward the given theory, he begins to impede that progress.

In the last part of this essay, let us consider (from the vantage point of historical epistemology) the question of the relative praxis-objective reference of W. V. Quine's comprehensive instrumentalism. How should this reference be characterized, given that Quine's conception after all is capable of implying a program of only limited functional scope with respect to theoretical break-through in science and of only limited instructive value for the research constituting this break--through ?

Let us observe, first of all, that the comprehensive instrumentalism of W. V. Quine inherits from positivist empistemology its individualistic orientation—a psychologistic version, moreover, of this orientation. This psychologism is behaviouristically oriented. The role of the 'immediate data' of traditional positivism is played, in the context of Quine's comprehensive instrumentalism, by the stimulus meaning of occasion sentences. That meaning, however, i.e., the class of concrete, physically characterized situations, in the context of which a given individual affirmatively reacts to the interrogative expression of a given occasion sentence, together with the class of other situations of this kind, in the contexts of which that individual reacts negatively to such a question, does not (in opposition to traditional positivist ideas) determine the logical value of related standing sentences in an unambiguous, 'natural', 'non-conventional' way. Additional premises, which are not determined empirically, are also necessary. These make up a system of ontological-semantic assumptions. These are linked, furthermore, to the stimulus meanings of particular occasion sentences in an individual way. In this, among other things, psychologistic individualism manifests itself in the methodology of Quine's epistemology.

Notice, however, that from the viewpoint of that epistemology, an essential distinction holds betwen the procedure for the selection of ontological-semantic assumptions by a particular individual beginning to master his own native language on the basis of the stimulus meanings of occasion sentences presented to him by his social environment, on the one hand, and the procedure realized by a scientific researcher when, for example, he learns a given ethnic language. (In order to maximize the instructive value of the example, let us assume that this is a language totally unknown to him until this time.) The difference arising between these two cases consists in the fact that in the first case, not only is the selection in question effected in a spontaneous way, a non-verbalized, non-conscious way, manifesting itself exclusively in the form of an appropriate set of dispositions toward particular verbalized behaviours, but it is also in principle the private affair of the given individual. The selection may be carried out in any manner at all, just as long as compatibility with the stimulus meanings of occasion sentences is maintained. It is only the stimulus meanings which can and must be uniformly recognized by the members of a given community. In the second case, however, the selection of ontological-semantic assumptions—a process that can be made conscious with variable success—is not only empirically limited by the requirement of compatibility with stimulus meanings, but by the additional requirement of its inter-subjective harmonization within a group of research specialists as well.

Of course, the distinction we have just noted—one which ought to be taken into consideration within the framework of the epistemology of language whenever the 'private' and/or scientific acquisition of language knowledge is examined— does not at all stand in opposition to the position of methodo- logical individualism. The fact that the development of scien- tific knowledge of a given language requires an inter-subjec- tive harmonization of the system of ontological-semantic as- sumptions imputed to that language only signifies, within the context of the entirety of outlooks of the author of *Ontological Relativity*, that one particular 'imputing' proposition which is legitimated by virtue of its optimal formal-technical qualities, i.e., which most 'economically' systematizes the empirical data registering stimulus meanings, 'carries the vote' within a group of specialists.

The discussion here is of the imputation, to be precise, of definite ontological-semantic frameworks to a given language since, as already been pointed out, the 'private' ontological- semantic decisions of a language's users, in the final analysis, are not decisions in the rigorous sense of the word : they are not the effect of some consciously directed mental processes taking place in reality. In accordance with his anti-mentalist position, W. V. Quine attributes real existence only to the ver- bal-behavioural dispositions themselves. Therefore, the 'pri- vately' operative system of ontological-semantic assumptions possesses only the status of a fictional, instrumental construct of the language researcher, and thus a wholly different status than that of its correspondence rendering which I have herein proposed : the methodological-theoretic humanistic coefficient of literal references. The system is not subject to hypothetical reconstruction. Concerning the meanings of words, the men- talism rejected by the author of *Ontological Relativity* assu- mes that 'words are supposed to be determinate in the native's mind, his mental museum, even in cases where behavioural cri- teria are powerless to discover them for us. When on the other hand we recognize with Dewey that 'meaning ... is primarily

a property of behaviour', we recognize that there are no meanings, nor likenesses nor distinctions of meaning, beyond what
are implicit in people's dispositions to overt behaviour.[11]

It is easily observed that the status of an instrumental construct which W. V. Quine ascribed to the imputed system of
ontological-semantic assumptions, to a system, that is, which
is mentally projected onto the whole of the verbal-behavioural
dispositions of individual users of a given language, precludes
even the very possibility of undertaking the explanative problem of the determinant of the shaping of that system. The
determinant of the formation of particular verbal-behavioural
dispositions can be invesigated, whereas this cannot be done
for the system of assumptions itself, since it is not a real object
for linguistic science research. Thus, it does not even make
sense in the context of Quine's position to pose the question of
whether this system of ontological-semantic assumptions is
defined by factors of a social nature or of an individual-psychological one. The matter presents itself differently in the case
of verbal-behavioural dispositions. They can and should be
explained in the spirit of individualistically oriented neobehaviourism.

We have so far been speaking of linguistic research, which—
according to W. V. Quine's point of view—can validly regard
verbal-behavioural dispositions as a real object of research, but
cannot so regard the instrumental construct called the system
of ontological-semantic assumptions. Notice that this position
illustrates, in a striking way, the marked inconsistency of the
author of *Ontological Relativity*. For this is no longer a comprehensively instrumentalist position, but in fact (like Carnap's) a dualistically instrumental one : the behavioural dispositions are real, whereas the system of ontological-semantic
assumptions has the status of a purely instrumental construction. This inconsistency, of course, is a consequence of the
acceptance of neobehaviourism—a psychological orientation
emphatically assuming this dualist instrumentalism.[12] As we
can see, therefore, W. V. Quine's solidarity with methodological

psychologistic individualism, which we have previously alluded
to, is so strong that he is prepared, in the name of that solidarity, even to retreat on certain points from consistent maintenance of a comprehensively instrumentalist viewpoint.

   Let us nevertheless call attention to the fact which the author of *Ontological Relativity* stated with regard to linguistic
research applies—according to his opinion, moreover—to all
epistemological research into science. For the theoretician of
scientific knowledge studying the language behaviour of representatives of particular sciences, bahaviour producing definite linguistic expressions, makes use of an instrumental construct in the form of a system of ontological-semantic assumptions as though these assumptions were consciously accepted
by those representatives. Just as in the case of linguistic research, the question of the determination of these assumptions
cannot seriously be posed in epistemology, much less the question of the determinant of the fact that the supposedly de-
tected—but in fact imputed by the theoretician of knowledge
acquisition—system of ontological-semantic assumptions of
a given set of scientific statements gained widespread acceptance among the specialiss of a given field. In this way, neo-
behaviourist methodological individualism has made it impossible to pose a question concerning the determinants—and these
are determinants of a social character—of the effective exchange of one system of ontological-semantic assumptions for
another. From this perspective, it can only be asserted—as
W. V. Quine has in fact done—that : (1) the representative of
a given scientific discipline, at the moment when he imputes
to a set of research results previously established within the
framework of that discipline a new system of ontological-semantic assumptions such that the real reflex of this course of
action is a change in the customary practices of acceptance and
refutation, a change which in particular reverses the logical
values of some observational sentences, acts as though in the
role of an epistemologist studying the language of those results ; and (2) his imputation has a chance of gaining general

acceptance only when it ensures a new, 'conomically' remun-
nerative conceptual organization of existing constructs; at
the same time averting any inconsistency which has revealed
itself between those constructs and certain newly noted observational data. From the gnosiological perspective of Quine's
instrumentalism, only in this way can the praxis of corresponding break-through (in an essentially corrective mode)
from the pre-theoretic stage of science be reconstructed and
validated.

It can easily be seen that this is not the gnosiological perspective of an active subject of the process of a science's transformations leading towards its theoretical stage. It is rather the
perspective of a passive participant in this process, a participant who—unlike the researcher of Carnap's perspective—
accepts the process and strives to rationalize it in the comprehensively instrumentalist manner available to him. Unquestionably, W. V. Quine's comprehensive instrumentalism—by virtue of the fact that it comprises an attempt to capture the specificity of the phenomenon of a theoretical system's formation—
also recognizes the process of individual assimilation of common
practical wisdom more adequately than does dualist instrumentalism, which 'gazes' at the same phenomenon from the
standpoint of a spokesman for pre-theoretic knowledge. For,
of course, both in the first and in the second case, we are dealing—in the light of the assumptions of methodological individualism common to both—with a description, the actual praxis-
-objective reference of which is not identical to the social
process of the formation of common practical wisdom, of pre-
-theoretic science, and eventually of theoretical science, but is
rather identical to the course of individual assimilation of the
knowledge constituting the result of each of these three processes, respectively. The point is that the course discussed above—
or to speak more precisely, its three respective cases—is more
adequately noted within the framework of Quine's conception
than it is from the perspective of dualist instrumentalism, because the viewpoint of an accepting registration of the fact of

the emergence of theoretical systems of knowledge, a view-point which seeks, moreover, something other than a pre-theoretic means of validating this fact, has rendered possible the discovery, although in an epistemologically distorted form, of the phenomenon of the functioning of the methodological--theoretic humanistic coefficient of the literal references of common-wisdom determinations and scientific ones as well. It is true, for reasons we have already discussed, that this dis-covery, the discovery of the system of ontological-semantic as-sumptions, was immediately down-graded by assigning this system a purely instrumental status. Nevertheless, it implies a certain essential correction (even if postulated only for formal--technical reasons) of Carnap's rendition (in the praxis-objective sense) of the process of the assimilation by individuals of com-mon practical wisdom. This correction follows from the recogni-tion of the fact that the mere sensory contact of an individual with the phenomena surrounding him does not unequivocally determine any conceptually articulated knowledge about the phenomena at all. Thas it, the individual's language behaviour, positively and negatively formulating appropriate states of af-fairs, cannot be explained until we ascribe to that individual (albeit in a purely instrumental manner) a definite system of ontological-semantic assumptions. Certainly, in order to notice that the system is determined socially and that, in a funda-mental way, it decides the shape of individual knowledge, one would have to give up basic tenets constituting the comprehen-sively instrumentalist orientation and most of all, those tenets which even conflict with the orientation itself : the tenets of psychologistic methodological individualism (in the neobe-haviourist version).

   In summary, therefore, we will say that while the relative (from the vantage point of historical epistemology) praxis--objective reference of Carnap's description of the opposition of theory and experience is the process of the individual assi-milation of scientific-theoretic concepts by a researcher who represents the pre-theoretic manner of thinking, the pre-theoretic methodological consciousness of science, the praxis–objective reference of Quine's comprehensively instrumentalist destruction of that opposition is the process of the individual assimilation of scientific-theoretic concepts by a researcher who passively participates in theoretic break-through : he accepts the results of break-through, but he is not the agent of those results. Furthermore, while the entirety of R. Carnap's epistemology comprises, as adequately as the positivist manner of thinking allows, an attempt at describing (in the praxis-objective sense of the term) the process of the assimilation by a given individual of the contents of common practical wisdom, of knowledge from the pre-theoretic stage and also of theoretical knowledge, the praxis-objective reference of the description embodied in W. V. Quine's epistemology is certainly an analogous process, but is conceived of in a way that casts in a different light some determinants of the process which are pointed out by positivism and signals some important additional determinants. Namely, determinants of the form of 'immediate data', which according to the positivists are supposed to suffice for the formation of that knowledge of the world which can be embraced within the (extended) observational language, do not emerge in an individual's field of vision in a spontaneous, 'natural' way, but as Quine assumes, are introduced into that field—in the form of stimulus meanings of occasion sentences—by the social environment of the given individual. Moreover, these data do not preordain any one particular crystallization of the 'observational' knowledge of the individual in question, since their conceptual articulation demands some additional activity on the part of that individual. The author of *Ontological Relativity* abandons this activity *via* the introduction of the instrumentally interpreted notion of a system of ontological--semantic assumptions. Indeed, such a conception makes it impossible to acknowledge either the high-priority determinational role of this system of assumptions or its social source. Nevertheless, this conception represents, in relation to the positivist description of the process of interest to us here, a gnosio-

160   THE DUHEM-QUINE THESIS

logical step forward that is sufficiently radical for us to regard Quine's understanding of the process as an essential corresponding correction of that description.

NOTES

[1] W. V. Quine, 'Two Dogmas of Empiricism', in : *From a Logical Point of View* ; 9 *Logico-Philosophical Essays*, Harvard University Press, Cambridge, 1964, p. 44.

[2] In addition to Carnap's dualistic instrumentalism and to comprehensive (total) instrumentalism, it is encumbent upon us to distinguish—as has been shown by K. Zamiara (*Epistemologia genetyczna a społeczny rozwój nauki (Genetic Epistemology and the Social Development of Science)*, Warszawa–Poznań, 1979)—instrumentalism associated with a psychological interpretation of Kantism, represented today by, among others, J. Piaget (and probably by C. Lévi-Strauss). This orientation assigns an instrumental status to all human knowledge with the single exception of psychological theory, which conceives to that knowledge as the effect and the tool of our adaptation to conditions imposed on the individual, located at a defined 'stage' of personal development, by the natural environment. This includes the social environment (which, however, is conceived of naturalistically).

[3] A particularly scrupulous presentation, in my opinion, of P. Duhem's outlooks concerning the question of interest to us is contained in Quine's article 'What Duhem Really Meant', in *The Boston Studies in the Philosophy of Science*, Vol. 14, Dordrecht, 1974 (pp. 37 and 46).

[4] Cf. Chapter 2.

[5] J. Giedymin, 'Odpowiedź' ('Reply') in *Teoria i doświadczenie (Theory and Experience)*, Warszawa, 1966, p. 165.

[6] *Ibid.*, p. 165. In a similar, 'generalizing' fashion. M. Hesse interprets Quine's reinforcement of Duhem's idea in her article 'Duhem, Quine and a New Empiricism', in *Can Theories Be Refuted ? Essays on the Duhem-Quine Thesis*, ed. S.G. Harding, Synthese Library, Vol. 81, Dordrecht–Boston, 1976. She explicates W. V. Quine's position in terms characterizing the mode of coding information by a computer in which coding is determined by conditions placed on the entirety of information, that which is being coded as well as that which has already been coded. These conditions determine the 'content' of individual bits of coded information (the 'content' of observational sentences) only indirectly and ambiguously. Certainly this latter point accurately expresses one of the most important aspects of W. V. Quine's concept. At this juncture, it is worth pointing out that the outlooks expressed by J. Giedymin at the *Teoria i doświadczenie (Theory and Experience)* conference (War-

szawa, 1964) together with the outlooks of his opponent, W. Mejbaum, concerning the question of interest to us here, comprise the 'last word' spoken on this topic in Polish philosophy of science.

7 A. Grünbaum, 'The Falsifiability of Theories ; Total or Partial ? A Contemporary Evaluation of the Duhem–Quine Thesis', in : *Boston Studies in the Philosophy of Science*, Vol. I, ed. M. W. Wartofeky, Dordrecht, 1963, pp. 179, 180.

8 *Ibid.*, p. 188.

9 *Ibid.*, p. 188. As we see, Grünbaum—in typical fashion assumes that : (1) at least some scientific statements (particular or general) belong to an observational language in the traditional sense that they are unambiguously 'naturally', 'unconventionally' (without the use of any additional assumptions) semantically interpreted in the domain of empirical data ; (2) if they were otherwise, all total theoretical systems would have an *a priori* character (the empirical conception of scientific theories can be maintained only upon assuming the existence of statements belonging to the observational language, as it is traditionally conceived of).

10 The author of *Ontological Relativity* himself says this, somewhat ironically, in a letter to A. Grünbaum, wherein 'for his own part', he writes that the thesis, as he uses it, is probably trivial. Cf. W. V. Quine, 'A Comment on Grünbaum's Claim, Can Theories be Refuted) ?', cited above, p. 132.

11 W. V. Quine, *Ontological Relativity and Other Essays*, New York–London, 1969, p. 29.

12 This state of affairs is convincingly demonstrated in K. Zamiara's book *Metodologiczne znaczenie sporu a status poznawczy teorii* (*The Methodological Significance of the Controversy over the Epistemological Status of Theories*), Warszawa, 1974, p. 98.

CHAPTER 5

ALTHUSSER'S INSTRUMENTALISM

5.1. A MARXIST VARIANT OF THEORETICAL HISTORICISM METHODOLOGY

The fact that the Marxist conception of historical cognition assumes a methodology of theoretical historicism[1] cannot be subject to doubt since (1) according to this conception, all studies of a course of societal events, and hence historical studies, ought to rely—given the explanation of particular social phenomena—on the theorems of historical materialism for their most general explanatory premises, and since (2) these theorems, as has been pointed out, have the character of nomological formulas.

Statements (1) and (2) together imply that the explanation of historical phenomena and also the laws that are used in this explanation have a particular character : a character specific to historical research.

It should be emphasized here that theoretical historicism methodology—as a norm of historical research—is not only a consequence of a conception of this research characteristic of Marxism, it is also a postulate advanced by the founders of Marxism that is specifically addressed to this research. This addressing, moreover, is understood very broadly, since at the very least 'all branches of science which are not natural sciences are historical.'[2] Hence, it is not simply methodological historicism that is proper to Marxism, but—as follows from the preceding remarks—a logically stronger position : theoretical historicism methodology in the field of all social sciences (humanities—in the broad understanding of the term that I have assumed). Incidentally, many data indicate that as a matter of fact, the norm of theoretical historicism methodology possesses,

[162]

in the intention of the classics of Marxism, a much vaster
scope of applications (transcending the domain of social sciences). This issue will not be taken up here, however.

On the other hand, the following point is quite essential for
the present discussion : theoretical historicism, like factographical historicism, is represented by an entire gamut of positions
of the most diverse types. The key to distinguishing between
particular variants of this version of historicism is provided
by analysis of the different manners of understanding the feature of the 'theoreticity' of knowledge. In using the term in
quotation marks in order to distinguish a group of positions
which—consequently—we qualified above with the name of
theoretical historicism methodology, we have taken one of the
broadest meanings of this term into account, a meaning which
is quite commonly associated with it. Namely, according to
this interpretation, the term signifies a type of knowledge
which is composed of certain general propositions, but is not
content with statements of single itemized facts. Nevertheless,
the expression 'theoretical knowledge' can be understood in
a variety of other ways. Unquestionably, when Karl Marx used
this very qualification, he had in mind much more than just
a set of general propositions, he had a special type of proposition in mind.

A fundamental Marxian intuition pertinent to this special
theoretical type of general proposition can be educed through
the analysis of, among other things, statements by the author
of Capital concerning classical political economy as well as
what he refers to as vulgar economics. The latter he evaluated
as a normal registration, organized in some way, of the common practical wisdom concerning management which the capitalist subject of production has at his command. The former, however, was viewed as an attempt at a theoretical conversion of that practical wisdom, an inconsistent attempt, and
therefore not entirely successful. The main idea of this evaluation can be conceived—more or less—as follows : as long
as we remain within the sphere of common practical wisdom

images of a given domain of reality, without seeking an answer
to the question of the objective mechanism lurking behind
these images and generating them, we find ourselves outside
the sphere of theoretical knowledge. We only know 'appearance';
we don't know 'essence'. Of course, 'appearance'—the picture of the world (of a definite domain) given to us by the wisdom of common sense, common practical wisdom—is not an
ordinary illusion or mistake. The practical effectiveness, however limited it may be, of the knowledge that draws the 'appearance' clearly indicates that in its own superficial way, this
knowledge reflects some real aspects of the objective conditions surrounding social praxis. Thus, disengagement from the
sphere of 'appearance', the transcendence of the framework of
the phenomenalistic knowledge embodied in common practical
wisdom, does not entail our simple rejection of this wisdom.
This transcendence requires that in constructing a theoretical
image of the objective mechanism generating common practical wisdom, we are able at the same time—if our image has
achieved this degree of adequacy—to answer the question of
how the existing practical knowledge had attained social acceptance (possibly limited by class). This is in fact an answer to
the question of the range of relative effectiveness in practice
of the existing knowledge, a range designated in terms of the
theory being built. For the fact that this range exists, that the
range is defined by the state of development of social praxis up
to the present, explains the situation that the knowledge of the
world we are now considering, a knowledge that understands
the world in the form of 'appearance', became socially (possibly
within the limits of a definite social class) accepted.

The relation holding between common practical wisdom and
the system of knowledge that explains its social acceptance
(in the way we have outlined above) has been identified in the
preceding chapters by the name of essentially corrective correspondence, while the corresponding system of knowledge has
been qualified as theoretical (in an absolute and also a relative
sense). 'Theoreticity' understood in this way appears to com-

prise a sufficiently precise rendering of the Marxian way in which the sense of this term (as well as—it seems—that of the term 'science') is conceived.

Thus, the Marxist concept of social research postulates something more than just theoretical (in the broadest sense of the word) historicism ; it postulates a historicism that is rigorously theoretical (i.e., it assumes 'theoreticity' of historical research in the sense that has just been expounded).

Distinguishing the Marxian, strictly theoretical variant of historicism from among the various versions of theoretical historicism permits a detailed contrasting of the latter with particular instances of positivist theoretical (in a broad sense) historicism. For the existence of specific historical regularities is also asserted within the framework of positivism. A closer inspection of the content of laws proposed in a positivist vein, for example, the laws of social evolution or laws elaborated in linguistics by the neo-grammarian school, makes it clear that we are dealing with certain phenomenalistic, common-sense generalizations having nothing in common with a strictly theoretical system of knowledge, with strictly theoretical, specifically historical laws. These generalizations, as a rule, aspire to the role of statements that belong to the so-called laws of temporal succession (of the type : 'Of two (similar) products of culture, that one arose later which has the more intricate structure', or 'The Proto-Slavic sound class of the shape TORT is transformed into a class of the shape TOROT in the East Slavic languages, a class of the shape TROT in the West Slavic languages and a class of the shape TRAT in the South Slavic languages', e.g. «worna»—«worona»—«wrona»—«wrana»). These laws, incidentally, recall the very similar general constructs of a natural science such as geology.

It has previously been stated that the norm of strictly theoretical historicism in the field of social sciences (humanities) is a consequence of the idea that the phenomena investigated by the social sciences ought to be clarified through reference to explanatory premises that are theorems of historical materialism or consequences deducible from those theorems. We can now make a still stronger statement.

The postulate of the building of strictly theoretical knowledge and hence—as we have seen—knowledge that captures the 'essence' lurking behind the 'appearance' described by phenomenalistic commonsense knowledge as embodied in social practical knowledge is also—as can be shown in detail—a consequence of the concept that historical materialism constitutes the explanatory foundation for research in the field of social sciences. For the thesis that the socially functioning subjective images of a given domain of reality are one thing, while the objective mechanism generating these images is something else, and the thesis that a theoretical description of that mechanism permits the fact of the social acceptance of these images to be explained are unquestionably a component of historical materialism. We therefore arrive at the following conclusion : the Marxist norm of strictly theoretical historicism methodology in the field of the social sciences (of the humanities) is equivalent to the norm that these sciences should assume historical materialism as their fundamental explanatory base.

The above conclusion is the logical strengthening, to which we have alluded, of the thesis which asserts that Marxist theoretical (in the strict sense of the word) historicism is a consequence of assuming historical materialism as the fundamental explanatory base of all social sciences. Nevertheless, the equivalence which our conclusion establishes still has to be somewhat reworded, since an understanding of the norm stated on the right-hand side of that conclusion is possible such that our equivalence would be invalid. It is possible to regard historical materialism as the fundamental explanatory base of social sciences in such a way that, at the same time, we do not in the least have to acknowledge the position of strictly theoretical historicism.

For example, L. Althusser, who undoubtedly accepts the norm on the right-hand side of the equivalence of interest

here, none the less states that 'Marxism is not historicism'.
Let us then consider the reasoning which induced this French
Marxist to advance such a thesis.

5.2. ALTHUSSER'S CONCEPTION OF HISTORICAL MATERIALISM

One of the chapters of *Reading 'Capital'* written by L. Althusser and—it is beyond doubt—accepted by the other author of
the bok, E. Balibar, bears a title evoking its central thesis:
'Marxism is not historicism'. Of course, the fact that Althusser
expressed one of his fundamental ideas by means of just such
a formulation does not necessarily mean that he is in particular an actual opponent of the use of the norm of strictly
theoretical historicism methodology in the domain of Marxist
research in the humanities. After all, the possibility that he
conceives of historicism in a different way than we have assumed here has to be considered. Let us therefore examine some
element of the argumentation employed by this French Marxist
for the purpose of substantiating the thesis of the anti-histori-
cism of Marxism. I seek to show that in fact, this thesis is intimately linked to the rejection of a realistic interpretation
of historical materialism, i.e., the rejection of the concept that
historical materialism represents an attempt at a theoretical
characterization of the real process of social development.

At least three factors make it clear that Althusser assumes
an instrumentalist (as opposed to a realist) interpretation of
historical materialism.

First of all, it is his view that 'while the production process
of a given real object, a given real-concrete totality (e.g., a
given historical nation) takes place entirely in the real and is
carried out according to the real order of real genesis (the
order of succession of the moments of historical genesis), the
production process of the object of knowledge takes place entirely in knowledge and is carried out according to a different
order, in which the thought categories which «reproduce» the

real categories do not occupy the same place as they do in the order of real historical genesis, but quite different places assigned them by their function in the production process of the object of knowledge'[3].

The above remark warrants commentary on several points. Yet we will limit ourselves here to the following statement. According to this French Marxist, the actual course taken by the process of social development is absolutely incomparable to the image of that process constructed by Marxist theory. For the actual process is shaped in the framework of a temporal sequence, whereas the theoretical image of the process is organized in a purely mental way. For this reason, the image cannot pretend to the role of reporting the actual state of affairs, and the theory that constructs this image thus cannot be interpreted realistically : as an attempt at reporting the actual state of affairs.

As I have indicated, in order not to break up the continuity of the train of thought, I want to refrain from formulating commentary at this point. Nevertheless, at least two questions demand our attention here, and I will refer back to them— directly or indirectly—somewhat later on : (1) Why is this colloquial image of history, shaped by pre-theoretical generalizations of common practical wisdom and expressing itself in phenomenalistic laws of temporal succession, considered by Althusser to be objective reality ? (2) Why does he identify the order of the mental process of constructing a theoretical image of history with the order postulated by that image (after all, in acknowledging an assertion of state of affairs A on the basis of the acknowledgment of an assertion of state of affairs B, or in some other way accepting a logical necessity to consider state B prior to state A, we make no presumptions about the nature of the substantive relationship linking these two states !) ?

Secondly, Althusser contrasts the image of history constructed by Marxist theory not only with objective reality, but also—and this is both easier to accept and more in agreement

with the remaining assumptions of the epistemology he profes-
ses—with the so-called 'empiricist' understanding of history. We
will later speak at greater length of the 'anti-empiricism' of
this French Marxist. The one statement of interest to us now
is that while the 'empiricist' understanding treats history as
something having arisen objectively, the Marxist theoretical
understanding is supposed to assume that the connections between particular fundamental links in social development, particular modes of production (from the primitive community to
socialism) are in fact only instrumental-theoretical mental
connections linking the ideas of these modes of production.
E. Balibar tries to characterize these connections by means of
a general 'theory of transitions' : a 'logical' connection between
the ideas of two successive modes of production is expressed
in the idea of the so-called structure of reproduction. Particular
variants of the structure of reproduction are expressed, it is
worth noting, by consecutive ideas of particular modes of
production.

Thus, this opposition in regard to the 'empiricist' image of
history also leads to an instrumentalist conception of the epistemological status of historical materialism, a conception which
is very accurately portrayed by two English Marxists, B. Hindess and P. Q. Hirst, in the following way : 'The theory of history that Althusser establishes is a theory of the distinct
eternities/totalities that are possible and a theory of the complex internal temporalities of these totalities ... History as a
time continuum and as a teleological process, in the real, is
superseded by history as a series of totalities considered as
eternities, within knowledge. These eternities have no necessary
succession, they do not form a historical progression, rather,
the eternities have a necessary connection within knowledge,
they form a logical series'[4]. As a matter of fact, then, we are
not dealing, in the case of Althusser, with an actual theory of
history. 'What is called a science of history is not a history'[5].
For in this conception, 'history is not a real object, an object
prior to and independent of thought'[6].

Thirdly and finally, Althusser's concept of theoretic (scientific) praxis as a field of operation detached from objective reality, existing wholly in the sphere of the purely mental, is not compatible with a realist interpretation of historical materialism as a theory characterizing the real development of the entirety of social praxis, scientific praxis included. By accepting the concept that 'the production of knowledge which is peculiar to theoretical practice constitutes a process that takes place entirely in thought'', Althusser in effect has no choice but to declare his support for an instrumentalist conceptualization of the Marxist theory of history.

If, for the authors of Reading 'Capital', historical materialism is not a realistically interpreted theory of historical social development, what is its role in relation to Marxist research into society ? This question is resolved in a clear manner by the above quoted authors, B. Hindess and P. Q. Hirst, who, although indeed engaging in polemics on several points with the French Marxists, fully agree with the instrumentalism of the latter. They also believe that 'development refers to the attributes of a structure designated by a concept and not to a ... process (itself).'⁸ Yet they preclude—unlike Althusser and Balibar—the existence of even 'logical' relationships between the ideas of particular modes of production. 'It is not supposed that more developed modes ... succeed less developed ones, or that there are any necessary relations of succession between modes of production ... The concepts of the modes of production ... do not form a history in thought, mirroring in their succession the evolution of the real.'⁹ Of course, the last sentence also expresses the outlook of the French Marxists, as does the following solution to the problem of the role of historical materialism. That is, it is not a theoretic characterization of the real course of history, but provides ideas aiding the construction of instrumental-theoretic conceptions of particular modes of production and especially of the so-called 'current situations'. 'These concepts are abstract, their value is not limited by the analysis of the concrete. As concepts they can have a theoretical function even if concrete conditions to which they are pertinent do not exist, have not existed and will not exist. Concepts which are not used in the analysis of concrete conditions are not therefore speculative and empty. It is empiricism which conceives the necessary field of application of all concepts as the real. In fact concepts have a valid field of application within theory"[10].

As we see, the instrumentalism of the authors under consideration here matches that version of the orientation which can be qualified as total instrumentalism, of the same type as W. V. Quine's (all facts stated by science have the character of instrumental-theoretic constructs). It is in opposition to dualist instrumentalism, represented by, for instance, R. Carnap (for whom there exist two kinds of facts that can be asserted by science : observable real facts and fictitious, instrumentally constructed theoretical facts). This opposition to dualist instrumentalism impels those authors as well as L. Althusser to reject Weber's conception of ideal types. For the latter is 'empiricist' : it treats the ideal model (type) as an instrumental theoretic construct, but confronts it with phenomenalistic data which (as is assumed within the framework of this conception) adequately describe objective reality.

Thus, from the standpoint of the outlooks under examination here, the Marxist theory of social development (a 'general theory of modes of production') cannot be understand as a set of general statements characterizing, in the manner of a framework, the regularities of real historical development, regularities in the light of which it would be possible to explain individual variants, also of a framework nature, characterizing respective formations and/or domains of the social developmental process. On the other hand, Marxist theory does constitute a certain kind of conceptual apparatus and therefore is represented by a system of methodological directives. These directives prescribe that the notions they distinguish be used in a specific manner in the building of appropriate instrumental--theoretic scientific constructs. Because the frequently voiced

slogan : 'Historical materialism—the methodology of the humanities !' (in conjunction with which, the understanding of
'methodology' as a system of directives is assumed) stands in
an unmistakable relation to the conception sketched herein,
this conception can be qualified as an understanding of historical materialism (exclusively) as a 'methodology' for research
in the humanities.

Thus, it has been illustrated that L. Althusser and the adherents of his viewpoint interpret historical materialism in a comprehensively instrumentalist fashion. Of course, this does not
eliminate the possibility of regarding the laws of social sciences, which assume the 'methodology' of historical materialism
and hence (in particular) are built with the help of framework
concepts provided by that theory, as laws of a specific nature.
However, it does preclude the possibility of acknowledging
these laws as strictly theoretical statements. For only when
we conceive of these statements in a realist manner can it be
asserted that they describe the real mechanism for generating
the essentially corrected, common sense generalizations of common practical wisdom overcome with their help. We see then
that the thesis on the subject of strictly theoretical Marxist
historicism formulated toward the end of the preceding section
of this essay essentially ought to be reworded in the following manner :

The Marxist norm of strictly theoretical historicism (methodology) in the social sciences (humanities) is equivalent to the
norm that these sciences should assume realistically interpreted
historical materialism as their fundamental explanatory
base.

5.3. 'ANTI-EMPIRICISM' AS A CONSEQUENCE OF THE 'METHODOLOGICALLY' INSTRUMENTALIST INTERPRETATION OF HISTORICAL MATERIALISM

It has already been affirmed that L. Althusser's exclusion of
the historical process of development of scientific practice

from the sphere of objective reality (this practice, i.e., 'production of scientific knowledge', is to take place exclusively in the mental sphere) can be reconciled only with an instrumentalist (and in particular, a 'methodological') interpretation of historical materialism. This interpretation can thus be regarded as 'responsible' for the fact that Althusser (as well as adherents of his conceptualization) understand scientific social practice in this way and no other.

I now hope to show that the general 'anti-empiricist' position assumed by the authors of Reading 'Capital' can in turn be considered a consequence of the isolation of science from socio-historical reality. At the same time, moreover, opposition to this isolation leads to a position compatible with 'anti-empiricism' only on certain points.

Let us now recall Althusser's characterization of 'empiricism'. It can be summarized in three points which express—according to this French Marxist—the fundamental assumptions of that orientation. This characterization assumes namely that : (1) to the subject of cognition is given a certain fragment of objective reality as the object of cognition, (2) the result of cognition is defined by features of that fragment of objective reality, and (3) cognition consists in the abstracting of essential features from the given fragment of objective reality. 'The empiricist concept of knowledge', writes Althusser, 'presents a process that takes place between a given object and a given subject ... The subject and object, which are given and hence pre-date the process of knowledge, already define a certain fundamental theoretical field, but one which cannot yet in this state be pronounced empiricist. What defines it as such is the nature of the process of knowledge, in other words, a certain relationship that defined knowledge as such, as a function of the real object of which it is said to be the knowledge.' ... 'To know,' we read further, 'is to abstract from the real object its essence, the possession of which by the subject is then called knowledge.'[11] For according to the empiricist, 'the real ... is structured as a dross of earth containing inside it a grain of pure gold, i.e., it is made of two real essences, the pure essence and the impure essence, the gold and the dross, or, if you like (Hegelian terms), the essential and the inessential'[12].

It may seem that to the extent to which we might be inclined to regard the 'empiricist' position as it is characterized by Althusser—and frequently formulated in fact in gnosiological reflection, moreover—to be inaccurate, the accuracy of 'anti-empiricism' ought to be acknowledged. This is not the case, however, since this 'anti-empiricism' takes a logically stronger stance than a mere negation of the 'empiricist' position : it is a negation of an extended consequence of 'empiricism'. According to this consequence, 'the object as well as the operation of knowledge ... are posed and thought (of) as belonging by right to the real structure of the real object'[13], while, as we know, for the 'anti-empiricist', not only are these distinct 'from the real object, knowledge of which (they) propose to produce'[14], but—what is more—both the 'object and the operation of knowledge' occur altogether outside of reality, as they belong to the sphere of thought.[15] This very assertion forms the fundamental premise of Althusser's 'anti-empiricism', and in particular of the obviously correct thesis that a given fragment of objective reality (the 'real object') is one thing, while its scientific image (the 'object of knowledge') is something else again.

This is not the place for a more precise treatment and evaluation of the degree of accuracy of the 'anti-empiricist' position. It is nevertheless worth noticing at this juncture that unto itself, the thesis of the variance between objective reality and the scientific (or colloquial) image of that reality, even if it is socially functioning, has to be acknowledged by everyone who does not identify, in an idealist manner, the objective world with the (individual-subjective or socio-subjective) image of the world. Yet the path by which L. Althusser arrived at this thesis may give rise to numerous reservations.

One observation which stands out sharply is that the con-

cept of the variance of the scientific picture of the world from
the objective world itself can be perfectly maintained without
having to locate the very process of producing that picture,
scientific social practice, in an ethereal, mental 'beyond'. It is
true that this practice forms its own pictures of reality or, to
be more precise, forms systems of knowledge, semantic models
of which in some way represent objective reality in the (socio-
subjective) framework of definite methodological-theoretic
assumptions which assign a certain (internal) 'order'—as is
emphasized by the authors of Reading 'Capital'—both to the
practice itself and to the picture of reality which that practice
produces. It is also true that this internal 'order' differs from
the 'order' of the objective historical development of scientific
social practice and that at best, it is only in intention isomorphic
to the 'order' which governs that domain of objective reality
of which social research praxis produces a scientific image.
But the fact that there exists such a thing as the sphere of
historically variable, methodological social consciousness, in the
light of which the society of researchers subjectively 'orders'
particular research actions in a certain way, the fact that there
exists such a thing as the historically variable sphere of socially accepted (at least within the community of scientific
researchers) research results, which assume the shape of respective 'ordered' theoretical systems and present a scientific
picture of the world that is 'ordered' in a specific way, does
not the least bit undermine the thesis that scientific social
praxis and its development are situated within the context
of the totality of real, historical social development. Nevertheless, one has to accept the proposition that the image of the
world produced by scientific social praxis—the semantic model
of a given system of scientific knowledge—is directly relativized in its content and in its 'order' to the appropriate methodological-theoretic humanistic coefficient, representing the
state of methodological social consciousness—which is given by
history and determined in a functional-genetic modality—as
well as the set of research results obtained up to the given

moment. However, the very use of the notion of social consciousness as a definite set of beliefs would surely be regarded
by Althusser—a 'theoretical anti-humanist'—as well as by his
followers[16] as evidence of concessions made towards 'bourgeois
idealist thought'.

There now arises a question of how the Althusserian 'anti-
empiricist', isolated from objective reality and lingering in the
sphere of the purely mental, is to go about substantiating his
claim that his theory is epistemologically valid. In *Reading
'Capital'* L. Althusser considers—as we might say—the realist
interpretation of this question, that is, the question of the relationship born by the research results produced by scientific
practice to objective reality, and he adopts a position, an entirely consistent one for him, of not attempting to answer the
question[17]. For in effect, as an 'anti-empiricist', he is unable
to respond to the question and thus does not even suspect that
his insightful assertion that the 'thought', in the domain of
which (we might say rather : in the subjective context of
which) social research practice takes place, 'does not work on
the real object but on the peculiar raw material, which constitutes, in the strict sense of the term, its «object» (of knowledge), and which, even in the most rudimentary forms of
knowledge, is distinct from the real object[18]' can be in harmony
with a realist answer to the question.

Whereas in *Reading 'Capital'*, Althusser does not advance
any concept of criteria for 'external' validation with regard to
norms and methodological directives that are functional in
science, some of his individual followers take a step further in
this respect : 'All Marxist theory', write the authors of the
book *Pre-capitalist Modes of Production*, 'however abstract
and general, exists to make possible the analysis of the current
situation ... The current situation does not exist independently of the political practice which constitutes it as an object.
The current situation exists for Marxist theory only in so far
as it is given a definite form by Marxist political practice, and
in so far as definite problems are designated as objects of

analysis or criticism within that practice. These problems are problems of political practice and are specified in political terms.'[19]

Hence, the measure of the epistemological validity of Marxist theoretical constructions is not their more or less directly assessed relation to objective reality, but their degree of usefulness for political practice, that is, the extent to which these constructions are helpful in the realization of current political tasks advanced by 'Marxist practice'. Of course, this kind of criterion for the epistemological validation of results of research into social phenomena can be successfully postulated chiefly with reference to research concerned with factors conditioning the course of political events. The point is, however, that that criterion can be located in the contexts of two different epistemological conceptions : (1) in the context of the assumption that the degree of accuracy of the cognitive rendering of the objective conditions of social praxis, including political praxis, determines the degree of the effectiveness of activity based on a theory that discerns these conditions, and (2) in the context of the 'anti-empiricist' assumption, according to which the relationship of the scientific image of objective reality to that reality is utterly irrelevant (or at least the role of this relationship cannot be determined), while the success of a political program derived from given theoretical solutions itself comprises a sufficient, as well as an indispensable, guarantee of the validity of those solutions. Conception (2) constitutes the result of the following train of thought : (a) the rejection of the norm of theoretical historicism methodology—(b) the instrumentalist interpretation of historical materialism—(c) the generalization of that interpretation to all scientific theories, a generalization appearing in the form of 'anti-empiricism'. The fourth element, in turn, of this train of thought voices the concept of the subordination of epistemological values to strictly momentary, extra-epistemological tasks, e.g., political tasks.

At this point, it is worth calling attention to the analogy

born by the conclusions reached in the field of the validity of
theoretic scientific constructs by the adherents of Althusserian
'anti-empiricism' to the concept of *verité à faire* propounded
by the Frankfurt School : research results achieved by the postulated 'hermeneutic-critical sciences' are in fact 'truths to be
realized' in practice, and not the more or less successful diag-
noses of an actual state of affairs.

The instrumentalism of the Frankfurt School fundamentally
concerns only the humanities, and not all sciences, as in the
case of 'anti-empiricism'. Also the point of departure for representatives of the Frankfurt School leading to 'revolutionary
Marxism', is not the 'anti-empiricist' conception, but rather
a variant of methodological anti-naturalism. Namely, they state
that the social, practical functioning of the humanities is
basically different from the practical functioning of the natural sciences. The latter function with respect to social practice 'in a technological manner', i.e., to put it briefly, they
provide (generally speaking, indirectly) practice with premises
of the type : if under circumstances $C_1, \ldots, C_n$, action A is
undertaken, result E will be obtained. On the other hand, the
humanities, 'hermeneutic-critical' sciences, function entirely
differently in the context of social practice. They do not fur-
nish 'technological' premises, but rather they, first, propose
definite values for acceptance and realization and, secondly,
'unmask' currently dominant values ('ideologies') by exposing
their genesis. The supposition that the humanities function socially in the same way as do 'empirical-analytic' sciences, i.e.,
'in a technological manner', and hence by providing schemata
for effective activity derived from theories that attempt to
discern objective reality, is considered to be an expression of
the positivist position. In conjunction with this, state the representatives of the Frankfurt School, the evaluation of the
theoretical works of Karl Marx emerges as ambiguous. As the
main creator of the conception of historical materialism, who
understands it as a realistically interpreted theory of social
development, comprising theorems on the basis of which acti-

vity can be undertaken to accelerate that development, he is a positivist. However, as the creator of the 'critical' theory of the capitalist modes of production, he can be fully accepted by the adherents of 'revolutionary Marxism'. For the theory, on the one hand, exposes the genesis of dominant 'ideologies' of the capitalist system in a critical manner, while on the other hand, it formulates certain *verités à faire*, advanced as values worthy of realization through 'emancipatory revolution-ary practice'.

Thus, a student of J. Habermas, Albrecht Wellmer writes : 'critical theory can prove itself only by initiating a reflective dissolution of false consciousness resulting in liberating praxis : the successful dissolution of false consciousness as an integra-tive aspect of emancipatory practice is the proper touchstone for its truth... The truth of critical social theory is a *verité à faire* ; in the last resort it can demonstrate its truthfulness only by successful liberation : hence the hypothetico-practical status peculiar to the theory. This hypothetico-practical status ... is challenged by the basic assumption of Marx's inter-pretation of history. In those basic assumptions, technical progress, the abrogation of «disfunctional» social repression and the dissolution of false consciousness are so indissolubly joined, that the irresistible advance of technical progress, which starts with the capitalist mode of production, has to be inter-preted as the irresistible advance towards the commonwealth of freedom.'[20]

Marx's 'mistake' lies in the fact that he did not understand that his critical theory (the theory of the capitalist mode of production) cannot be understood after the pattern of 'strict experimental scientific theories'. For if it were, it could not then 'anticipate the necessary transformation of men who want to transform society, in the dimension of a self-enlightening praxis'. As a result, the author of *Capital* 'must rigorously distinguish between the transformation of consciousness and the transformation of attitudes and modes of behaviour : the first becomes knowledge of the economic law of movement of capitalist society, the second must be comprehended as the necessary result of the process of material production'[21].

This is the outcome of 'the latent positivism of Marx's philosophy of history'[22], that is, of the outlook that historical materialism accounts for real regularities of social development, knowledge of which constitutes the point of departure for the development of the revolutionary praxis of the proletariat. Marx's supposition that any regularities of this type exist whatsoever supposedly stems, in turn, from his lack of awareness of the existence of this basic difference between 'empirical-analytical' sciences, which discover regularities that are recorded in the form of general statements composing premises for technological 'prescriptions', and 'hermeneutic-critical' sciences which do not provide such 'prescriptions', but only formulate *verités à faire*. From the vantage point of these postulated truths, they (the 'hermeneutic' sciences, that is) adaptively interpret even the past. For Marx is supposed to have mistakenly believed that all social praxis and, hence, the revolutionary praxis of the proletariat in particular, has the structure of material productive praxis—requiring technological premises.[23]

It is not my intention to erase the differences appearing between Althusser's epistemology and the methodological concepts of the Frankfurt School. I am only seeking to point out the convergence that arises in the two following points. First of all this convergence is manifested in the realm of the manner of understanding the relationship of scientific praxis, in the first case, or research in the humanities and the 'Marxist revolutionary praxis' connected therewith, in the second case, to objective reality. In both cases, the relevant praxis is located outside of that reality, in the subjective, mental sphere. In both cases, an opposition with the praxis of material production is delineated. Secondly, the issue of the epistemological validity of the products of scientific praxis, in the first case, either is left unresolved or is regarded as a question of the degree of usefulness of these products of science to the tasks of the political praxis of the proletariat. In the second case, this latter possibility is also exploited for solving the problem of the validity of research results in the humanities : it is the aim of the 'Marxist revolutionary praxis' of the proletariat aspiring to liberation that such and such states of affairs should become facts, and to the extent that this praxis achieves success, the formulations of these facts attain the status of scientific theorems.

This intellectual ultra-revolutionism, rejecting a realist ('historical', 'empiricist', 'positivist') interpretation of historical materialism, has two distinct mental contexts, however. In the first case, we are dealing with 'anti-empiricism', which conceives of all products of scientific research instrumentally. 'Anti-empiricism' is unavoidable, given an instrumentalist (and in particular, 'methodological') interpretation of historical materialism. In the second case, we are dealing with methodological anti-naturalism, which in a radical way distinguishes the natural ('empirical-analytic') sciences from the humanities ('hermeneutic-critical sciences'). This distinction is based on the supposedly diametric opposition of the two separate manners of practical functioning of the first and second groups of sciences. This opposition is linked to the assumption that the material praxis of production, requiring 'technological' premises provided by the 'empirical-analytical' sciences is one thing, while the emancipatory praxis served by the ('hermeneutic-critical') humanities, which produce *verités à faire*, is something else again entirely. It follows, therefore, that natural sciences function in social praxis by providing knowledge about means of realization of definite states of affairs, whereas the humanities function by providing world-view systems or ideological (in the 'non-Frankfurt' sense of the word) systems.

Let us note two more observations. Firstly, it is worth emphasizing here that the epistemological conceptions of the Frankfurt School constitute a particularly striking confirmation of the thesis of the clear-cut relation of methodological anti-naturalism to an assumption which can easily be interpreted as a thesis that the task of the humanities is exclusively the world-view valorization of immediate-practical values[24]. Secondly, notice that Althusserian epistemology—although it also leads (by means of instrumentalism) to the subordination of gnosiological purposes to world-view ideological purposes alone —remains, from a certain perspective, on the opposite pole : methodological naturalism is assumed here in the form of an extremely radical thesis of 'theoretical antihumanism'.

NOTES

[1] Cf. the first chapter.
[2] F. Engels, 'Karl Marx—A Contribution to the Critique of Political Economy', in: *K. Marx, A Contribution to the Critique of Political Economy*, Moscow–New York, 1970, p. 220.
[3] L. Althusser and E. Balibar, *Reading 'Capital'*, translated by B. Brewster, New York, 1970, p. 41.
[4] B. Hindess and P. Q. Hirst, *Pre-capitalist Modes of Production*, London 1975, p. 316.
[5] *Ibid.*, p. 317.
[6] *Ibid.*, p. 318.
[7] L. Althusser, E. Balibar, cited above ; p. 42.
[8] B. Hindess, P. Q. Hirst, cited above, p. 321.
[9] *Ibid.*, p. 321.
[10] *Ibid.*, p. 321.
[11] L. Althusser and E. Balibar, cited above, pp. 35–36.
[12] *Ibid.*, p. 36.
[13] *Ibid.*, p. 39.
[14] *Ibid.*, p. 39.
[15] Cf. footnotes 3 and 7.
[16] I have in mind here in particular some of those who champion the slogan : 'Historical materialism—the methodology of research in the humanities'.
[17] Cf. L. Althusser and E. Balibar, cited above, where we read (p. 61) 'The reader will understand that I can only claim, with the most explicit reservations, to give the first arguments towards a sharpening of the question we have posed, and not an answer to it', and again later (p. 68) 'I shall leave the question in this last form, and merely recall its terms'.

[18] *Ibid.*, p. 43.

[19] B. Hindess and P. Q. Hirst, cited above, p. 322. It should be pointed out that since the time of *Reading 'Capital'*, Althusser has gradually drawn closer to the position represented here. This can be seen, for instance, in his *Réponse à John Lewis*, Paris, 1973.

[20] A. Wellmer, *The Critical Theory of Society*, New York, 1971, pp. 72–73.

[21] *Ibid.*, p. 74.

[22] *Ibid.*, p. 67–119.

[23] This very opinion is the premise of the outlook voiced by Habermas that the mental source of the 'positivist distortions' appearing in the theoretical works of Karl Marx is his 'equation' of the notion of material-productive labour with the notion of social praxis in general. This 'equation' demands analogous subjective-rational premises for every type of praxis, and therefore, in particular, premises suited to 'technological' functioning. Cf. in this matter, J. Habermas, *Erkenntnis und Interesse*, Frankfurt, 1968, p. 153.

[24] I advance this thesis in *Szkice z teorii poznania naukowego* (*Essays on the Theory of Scientific Knowledge*), Warszawa, 1976.

SUBJECT INDEX